SUGARCANE PRODUCTION

ADVANCE TECHNIQUES

The Authors

Dr. Lalita Rana is post graduated and Ph.D from Bidhan Chandra Krishi Viswavidhyalaya, Mohanpur, West Bengal. She has published a good number of research papers in National and International Journals and book chapters aid value to her name. She is actively participated in various activities like teaching of M.Sc. and Ph.D students and guidance of M.Sc. (Ag.) students Presently, she is working as Assistant Professor in the Department of Agronomy under Dr. Rajendra Prasad Central Agricultural University, Pusa, Samastipur, Bihar.

Dr. Anil Kumar is presently working as Assistant Professor-cum-Scientist since July, 2007 at SRI, Department of Entomology, RPCAU, Pusa, Samastipur (Bihar). He did his B.Sc. (Ag.) and M.Sc. (Ag.) in Entomology and Agril. Zoology at BHU, Varanasi (U.P.) and Ph.D. at Uttar Banga Krishi Viswavidalaya, Pundabari, Cooch Behar (W.B.). He has published more than 40 research articles, 25 popular/technical articles and 10 book chapter. He has also published 04 edited and 03 authors book

Mr. Manish Kumar is graduated in Agriculture from Banaras Hindu University, Varanasi, Uttar Pradesh in 2012 and post graduated in 2014. He has published number of research papers, popular articles and book chapters. Presently, he is working as Agriculture field officer and also involve in extension activities in field of agriculture.

Dr. Navnit Kumar obtained Graduation, Masters and Ph.D. degree from Rajendra Agricultural University, Pusa, Samastipur, Bihar. He has an experience for over 13 years in teaching, research and extension. He guided seven students of M.Sc. (Ag.) majoring in Agronomy and published 50 research papers in national and international journals with 50 popular articles in farmers friendly magazines.

Dr. Sudhanandini is presently working as Assistant Professor-cum-Scientist since Feb, 2019 at Department of plant Pathology, TCA, Dholi, RPCAU. She did her M.Sc. (Ag.) and Ph.D. at GBPUAT, Pantnagar (Uttarakhand).

Dr. Anil Kumar Singh is presently working as Director, Sugarcane Research Institute, RPCAU, Pusa Samastipur, Bihar. He has completed Graduation, Master's and Ph.D. degree from IAS, Banaras Hindu University, Varanasi, Uttar Pradesh. He is the recipient of Rajiv Gandhi Gyan Vigyan National Award 2012 by Hon'ble President of India. He published more than 100 research papers in National and International Journals and also written many book chapters, popular/technical article and authored/edited book to his credit.

SUGARCANE PRODUCTION
ADVANCE TECHNIQUES

Lalita Rana
Assistant Professor
Sugarcane Research Institute,
Dr. Rajendra Prasad Central Agricultural University,
Pusa, District Samastipur (Bihar)

Anil Kumar
Assistant Professor
Sugarcane Research Institute,
Dr. Rajendra Prasad Central Agricultural University,
Pusa, District Samastipur (Bihar)

Manish Kumar
Agriculture Officer, Bank of Baroda
Nawabganj, Allahabad (Uttar Pradesh)

Navnit Kumar
Assistant Professor
Sugarcane Research Institute,
Dr. Rajendra Prasad Central Agricultural University,
Pusa, District Samastipur (Bihar)

Sudhanandini
Assistant Professor
Tirhut College of Agriculture,
Dr. Rajendra Prasad Central Agricultural University,
Pusa, District Samastipur (Bihar)

Anil Kumar Singh
Director
Sugarcane Research Institute,
Dr. Rajendra Prasad Central Agricultural University,
Pusa, District Samastipur (Bihar)

2022

Daya Publishing House®
A Division of
Astral International Pvt. Ltd.
New Delhi – 110 002

Cataloging in Publication Data--DK
Courtesy: D.K. Agencies (P) Ltd. <docinfo@dkagencies.com>

Rana, Lalita, author.
Sugarcane production : advance techniques / Dr. Lalita Rana [and five others].
pages cm
ISBN 9789390435043 (HB)

1. Sugarcane--India. 2. Sugarcane industry--India--Management. 3. Sugarcane products--India. I. Title.

LCC HD9116.I4R36 2021 | DDC 338.133610954 23

Published by : **Daya Publishing House®**
A Division of
Astral International Pvt. Ltd.
– ISO 9001:2015 Certified Company –
4736/23, Ansari Road, Darya Ganj
New Delhi-110 002
Ph. 011-43549197, 23278134
e-mail: info@astralint.com
Website: www.astralint.com

Preface

The Ever-evolving population and stagnating yield potential of sugar crops forcing the scientists, researchers, traders and policymakers to develop appropriate field management (agronomic practices) of sugarcane. As sugarcane is a good source of sugar as well as other biofuels it has the potential for increasing foreign exchange. The crop has not only been contributing to the growth of the sugar industry but also for alcohol production for petrol blending, thermal power generation and charcoal industry, etc. so it has great potential for achieving maximum Gross Domestic Product (GDP) in the national economy. Sugarcane area in India was 5.06 million hectare and production of about 405.42 million tone with average productivity of 80.10 t/ha during 2018-19.

The Present book is an effort in this direction in which the author has attempted to describe the recent advance techniques with regards to sugarcane crop. The various aspects of the crop cultivation related to the origin and history, area and distribution of sugarcane in India as well as the world, Botanical distribution, varieties having different traits and suitable for climatic and soil conditions, cropping systems based on sugarcane crop, seed setts and planting methods, equipment and implements for sugarcane cultivation, nutrient management (organic and inorganic approach), water management, weed management, management of sugarcane diseases and insect pests, harvesting including a mechanized way of ratoon management and devices. Description of planting methods, modern approach of nutrient management, mechanization finds an important place in the chapters. Different intercropping systems with autumn and spring-planted sugarcane have been mentioned which could be beneficial for fetching more income for sugarcane growers. In respect of weed control, pest and disease control, the outward signs of the disease or attack of insects-pests, their mode of spreading, modern and effective control measures get prominence in descriptions. All the latest research findings in various aspects of crop production have been incorporated in the text. The book contains informative, elaborative, authoritative and readily usable information on the improved agronomic practices for a step forward the crop productivity. The book is written in simple language which is easy to understand. This book will be helpful for students, researches, sugarcane growers.

I greatly acknowledge all authors who contributed to writing this book. I also acknowledge the various resources, reference books and research papers from which valuable materials for this book were drawn. I am deeply indebted to the entire staff of the Department of Agronomy, Entomology, Pathology and other teaching and non-teaching staff of Sugarcane Research Institute, Dr. Rajendra Prasad Central Agricultural University, Pusa, Samastipur, and Bihar for the appreciation rendered by them in the preparation of this Book. I express my sincere gratitude to my Advisor Dr. Hirak Banerjee, Associate Professor, Agronomy, Bidhan Chandra Krishi Viswavidhyalaya who inspired me to write this book and gave valuable suggestions from time to time.

Last but not least I also thank to my husband Arun Singh Rana, and my colleagues for providing encouragement to complete this work.

Lalita Rana

Anil Kumar

Manish Kumar

Navnit Kumar

Sudhanandini

Anil Kumar Singh

Contents

Chapter

1

INTRODUCTION

INTRODUCTION

Sugarcane (*Saccharum officinarum L.*) is cash cum industrial sugar crop in India, which holds a prominent position in the national economy. Being a C_4 plant it play an important role in Carbon dioxide (CO_2) fixation and lowers down the photorespiration activity. Sugarcane is a tall, erect perennial plant growing upto 5-6 meters with multiple stems. It is the long duration sugar crop which is cultivated in an area of about 5.06 million hectares with an annual production of 405.42 million tones. India has the largest area under sugarcane cultivation and globally, India stands second in sugarcane production after Brazil. Sugarcane is grown worldwide, between the latitude 36.7° N and 31.0° S ranging from tropical to subtropical areas. Approximately, 80 per cent of sugar is derived from sugar cane and the remaining 20 per cent is derived by sugarbeet. It offers job to over a million people, directly or indirectly, in addition to making a substantial contribution to the national exchange. It is primarily grown as a sugar crop. Sugar juice is used for making sugar, khandsari and jaggery (gur) and sugarcane by-products like bagasse, molasses, etc. Now a days, used for production and manufacture of ethanol and energy production in commercial scale. In Sugarcane excess number of tillers produced which is not desirable for juice quality, so these excess tillers can be used for fodder purpose. Bagasse is generally used for fuel purposes, but now it can be used for making cutlery which is biodegradable and eco-friendly. Press mud is used for manuring in alkaline and saline soils. Sugarcane gaining more importance in the current scenario and Government is also promoting sugarcane production on a large scale in the country. Agricultural policy in India across decades has been focusing on self-reliance in food grain which we achieved through Green Revolution. The national economy of India is heading for an inclusive growth and in making this progress a reality livelihood generation from core agricultural and rural sector must have to go all out and sustainable livelihood generation mentoring customization and value addition have got both intrinsic and extrinsic dynamism.

1.1. Origin and history

Cultivation of sugarcane in India dates back to the Vedic period. Charaka wrote at the beginning of the Catholic period the oldest treatise on Hindu medicine. Throughout this work, the glossary of Sanskrit denotes an extensive knowledge of the different types and preparations of sugar that is considerably more than 2000 years old. The word is derived from the Sanskrit word 'Sakkara' or 'Sarkara'. Sanskrit literature from ancient India, written about 1500-500 BC, offers the first record of the cultivation of sugar cane and the manufacture of sugar in the Bengal region of the Indian subcontinent. The Sanskrit term for a crudely manufactured sugar substance was guda, meaning "to create a ball or conglomerate."

Extraction of sugar cane juice from the sugarcane plant, and eventual domestication of the plant sometime around 4000 BC in tropical South East Asia. The discovery of the manufacture of cane sugar granules from sugarcane juice in India over 2000 years ago followed by developments in the processing of crystal granules in India. There are two domestication centers for sugarcane: one for Papuans in New Guinea *Saccharum officinarum*, and another for Austronesians in Taiwan and southern China *Saccharum sinense*. Papuans and Austronesians initially used sugar cane mainly as food for domesticated pigs. Spreading the two *S. Bureau* and *S. Sinense* is closely related to the Austronesian people's migrations.

Saccharum officinarum was initially domesticated by Papuans in New Guinea and the islands east of the Wallace Line where it is the current center of diversity. They were selectively bred from the native *Saccharum robustum* beginning at around 6,000 BP. After interaction with the Austronesians, it spread westward from New Guinea to Island Southeast Asia where it hybridized with *Saccharum spontaneum*.

1.2. Area and Distribution

The world's sugarcane growing countries lie between 36.7° north latitude and 31.0° south of the equator extending from tropical to subtropical areas. Major sugarcane growing countries are Brazil, India, Cuba, Ghana, etc.

Uttar Pradesh, is the highest sugar-producing state in the country, is estimated to have a sugarcane area at 2.21 million hectares, as against 2.22 million hectares in 2018-19 SS i.e. a marginal decrease of about 2% with high numbers of sugar mills. Maharashtra's cane area has gone down by about 30% for 2019-20 SS, mainly due to poor rainfall from September'2018 onwards, followed by low reservoir levels, which adversely affected sowing of 15 months and 12 months crops.

Sugarcane is cultivated in about 500 districts spread across 26 states. But 161 districts spread over 10 states contribute over 95 per cent of the area and production and are designated 'dominant sugarcane producing districts'. The area (2.22 mha) and production (179.71 million tonnes) under sugarcane was highest in Uttar Pradesh but average productivity was highest in Tamil Nadu (103 t/ha) during 2018-19.

Table 1: Top five Major sugarcane growing countries

S.No	Country	Area (Million ha)	Production (Million tonnes)	Yield (Tonne/ha)	Sugar Production (Million tonnes)
1.	Brazil	9.83	739.27	75.17	35.80
2.	India	5.06	341.20	67.43	27.25
3.	China	1.82	125.54	69.01	13.30
4.	Thailand	1.32	100.09	75.74	10.20
5.	Pakistan	1.13	56.48	63.75	4.70

Source: India stat, 2018-19

Table 2: Area, Production and Productivity of sugarcane growing states (2018-19)

S.No.	States	Area (million ha)	Production (million tonnes)	Productivity (t/ha)
1.	Uttar Pradesh	2.22	179.71	80.81
2.	Maharashtra	1.16	89.77	77.20
3.	Karnataka	0.47	42.41	90.00
4.	Bihar	0.23	20.12	89.18
5.	Gujarat	0.15	11.33	73.18
6.	Tamil Nadu	0.17	17.14	10.3.00
7.	Haryana	0.11	8.51	78.24
8.	Andhra Pradesh	0.10	8.09	79.36
9.	Madhya Pradesh	0.11	5.28	48.91
10.	Punjab	0.10	7.77	81.83

Source: India stat, 2018-19

1.3. Sugarcane growing regions in India

1.3.1. Tropical region

Tropical region comprises two major agro-climatic zones viz., Peninsular zone (Maharashtra, Tamil Nadu, Karnataka, Kerala, Some parts of Andhra Pradesh Gujarat and Madhya Pradesh) and east coast zone (coastal areas of odisha, Andhra Pradesh, Tamil Nadu).

Productivity of sugarcane is higher (80 t/ha) in the tropical regions than subtropical regions (60 t/ha). In the tropical region, sugarcane gets more or less ideal climatic conditions for its growth and development throughout year. Higher sugar recovery due to long sunshine hours, cool nights with clear sky and the latitudinal position of the area help in higher sugar accumulation in plant. Although sometimes biotic and abiotic stresses faced by sugarcane crop during tillering and grand growth phases become critical to sugarcane crop and affect growth and yield adversely.

1.3.2. Sub-tropical region

The subtropical region comprises three sugarcane agro-climatic zones like North-West Zone, North-Central Zone and North-East Zone.

1. **North-West Zone:** Western Uttar Pradesh, Punjab, Haryana, Western Rajashan comes under N-W zone. In this zone, temperature is very low during the month of December-January. Due to low temperature in this region, growth of sugarcane is restricted for 3-4 months.
2. **North-Central Zone:** Uttar Pradesh (U.P), Bihar, Haryana and Punjab states face the extremes of climate viz., high and low temperatures, relative humidity, sunshine hours and wind velocity etc. Sugarcane production and productivity in this region is very much influenced by the climatic conditions throughout the year. Uttar Pradesh is having maximum area under sugarcane cultivation. However, the highest sugar recovery can be obtained in the Maharashtra. Climatic conditions are generally variable depending upon the seasons and sometimes within the season also. Sugarcane crop faces all the seasons in a year.
3. **North East Zone:** It comprises Eastern U.P, Bihar, and West Bengal, sugarcane suffers due to floods and water logging during monsoon months. The high incidence of pests and diseases is major bottle neck in achieving higher sugarcane production. The cane yields are lower in the subtropics due to short growing season, moisture stress, more pest and disease problems, floods and water logging and very poor ratoons. The average yield of this region is around 60 tonnes per hectare.

Chapter

2

CLASSIFICATION AND BOTANICAL DESCRIPTIONS OF SUGARCANE

2.1. Classification

Sugarcane is classified into 3 cultivated and 2 wild species. They are as follows.

2.1.1. Cultivated species

1. *Saccharum officinarum* Linn.
2. *Saccharum barberi* Jesw.
3. *Saccharum sinense (Roxb)* Jesw.

2.1.2. Wild species

1. *Saccharum spontaneum* Linn.
2. *Saccharum robustum* Brand and Jesw.

Saccahrum officinarum (2n = 80) is known as noble cane and it is grown for chewing purposes having a higher percentage of sucrose and a lower percentage of fiber than other varieties. This species has thick culms in various colors with broad leaves. It consists of a big inflorescence.

S. barberi (2n = 82 to 124) the species is characterized by a thin cane of North India. It consists of thin stalks with narrow or medium leaves. The inflorescence is small. It is well adapted to the sub-tropical region and tolerant to insect pests and diseases.

S. sinense (Roxb) Jesw. (2n = 118) This species is introduced from south east China into India consists of thicker and softer with broader leaves. It is common species of North India having symmetrical branching habits. The inner canes are straight and outer ones are curved, thus the culm shows a cup-shaped appearance. The Internodes are long with prominent nodes. The clones are susceptible to moisture stress and red rot diseases.

S. spontaneum L. (2n=40 to 128) these are wild species, small and bushy which are variable in forms from thin to thick types. It is resistant to waterlogged conditions and used for the breeding purpose for developing resistance to insect pests and diseases.

S. robustum Brand and Jesw (2n= 60 to 80) the species are wild having strolling habit with high fiber content. The stalks are thick

2.2. Botanical descriptions

Sugarcane belongs to the Gramineae family. It is a tall perennial grass of 2-7m height having tillers at the base for producing un-branched stems. The plant is composed of four principal parts, the root system, the shoot system, leaves and inflorescence. The description of these parts is given below.

2.2.1. Root system

Sugarcane has a fibrous root system. According to their origin, they may be classified as sett roots and shoot roots.

2.2.1.1. Sett roots

Sett roots are temporary roots that develop from the dormant roots of the root zone. These are thin and much-branched and functions for a limited period. These roots provide moisture and nutrients for the growing primary shoot until if form roots of its own. Later on, these 'sett roots' cease to function and die.

2.2.1.2. Shoot roots

Shoot roots are developed from the shoots base after the buds sprout and sett roots disappear after the shoot roots grow. They are permanent dense, fleshy, and white tips. There is continuous development of new roots from tillers.

2.2.2. Shoot system (stalk)

Sugarcane stalk is roughly cylindrical and having nodes and internodes. It is an above-ground portion of the plant which bears leaves and flowers. Each node consists of a bud also known as eye arranged in an alternate fashion. These buds are protected by leaf sheath, leaf scar is found just below the bud. Waxy materials are present in the internodes of some varieties. Root band consists of translucent dot-like structures known as root primordia and give rise to sett roots. The underground portion of the stem tapers rapidly and from the lateral buds of this region, shoots develop called tillers.

2.2.3. Leaves

The sugarcane leaves consists of leaf blade (lamina) and the leaf sheath. The leaf sheath has certain characters like spines or hairs on the back of the sheath, ligule (a lateral extension of the sheath/ paper-like structure, auricle. The lamina is long, a flat structure usually 5-6 cm wide, 1to 1.5 m long. The midrib is usually whitish but sometimes red or purple color is also found. The leaf edges are generally serrate.

2.2.4. Inflorescence

The inflorescence of sugarcane is commonly known as "Panicle" or "Arrow". It is about 25-30cm long and tapering. The arrangement of the spikelets is racemose i.e. the oldest flowers at the bottom and the youngest at the top. The spikelets occur in pairs, one of which is pedicellate or stalked and the other sessile. At the base of each spikelet, there is a tuft of long silky hairs. Many seeds are infertile, they are usually small and of low vitality. Flowering in sugarcane occurs at an age of 300-350 days in case of south Indian conditions and in North Indian conditions, do not flower at all.

Now a day, breeders of different institutes, universities, etc. going through the research for the production of sugarcane by seeds. However, plants grown through seeds are not vigorous but still inflorescence of sugarcane is important for breeder seed production.

Chapter

3

CROP GROWTH STAGES AND METEOROLOGICAL REQUIREMENTS

3.1. Crop Growth Stages

3.1.1. Germination stage

The germination stage is very important for the maintenance of the plant population to ensure a better yield of sugarcane. The atmospheric temperature, soil moisture and aeration are the most important factors affecting germination. Germination of sugarcane setts is very slow when the temperature goes below 20° C and that above 20° C, it increases progressively to an optimum between 80 and 90°F. For good germination, there should be sufficient moisture in the soil. Lack of adequate moisture greatly influences the germination.

3.1.2. Tillering stage (Formative stage)

When the plant germinates, it forms short joints under the soil. Each joint has got a bud and thus the basal part provides a large number of buds. These buds further sprout and give rise to the tillers which developed into secondary and tertiary canes. Barber studied the mode of tillering. The development of a tiller depends on time, angle and depth of emergence. The angle of emergence determines the type of tillering. There are four types of tillering. The type of tillering is very important in governing crop yield.

3.1.2.1. Cup-shaped tillering

This type of tiller is generally found in those varieties which grow usually upright. They are usually deep-rooted and grow rather slowly during the early period of tillering. The clump in such types is compact and dense and results in homogenous development of almost all tillers.

3.1.2.2. Bowl-shaped tillering

Both mother stalk and first formed tiller posses slanting position, with 40 to 60° angle and sometimes 70 to 80° with the ground surface. The stool of this type of tillering covers more space than the cup-shaped tillering.

3.1.2.3. Dish shaped tillering

The acute angle from 0 to 25° with ground surface formed in both the mother stalk and the first tiller. The characteristic of this type of tillering is the sprawling growth habit. Varieties of this type of tillering usually have shallow roots.

3.1.2.4. Fan shaped tillering

The mother stalk of the stool is usually erect while the lowest two primary tillers emerge from the ground surface with an angle of 40° to 70°. Till the middle of the tillering period, it remains fan-shaped but later on, it gets modified with the emergence of the secondary tillers. Which generally emerge later than in other types of tillering.

3.1.3. Grand growth stage

The Grand growth period starts 120 days after planting and lasts in a 12-month crop up to 270 days. The stabilization of this phase tiller takes place during the early stage. Out of the total tillers produced, only 40-50 percent of the survived by 150 days to form a milliable cane. This is the most vital growing cycle of the crop in which the actual cane formation and elongation occurs, and thus yields build-up. It is the crop's most significant growth step in which the actual cane formation and elongation take place and hence yield build-up. During this process, leaf development is frequent and rapid with leaf area index reaching 6-7. In favorable conditions, stalks grow rapidly almost 4-5 internodes per month. Warm, humid and sunny climatic conditions favor better cane elongation. Moisture stress reduced intermodal length. A temperature around 30°C with humidity of around 80 % is most conducive to good growth.

3.1.4. Maturity stage

The maturation and ripening period in a 12-month crop lasts from 270-360 days for about three months. Synthesis of sugar and rapid accumulation of sugar occur during this process, and vegetative growth is reduced. When ripening progresses, simple sugars (monosaccharide viz. glucose and fructose) are converted into disaccharide as cane sugar (sucrose). Cane ripening proceeds from bottom to top part contains more sugar than top portions.

3.2 Meteorological requirements

Sugarcane is a tropical plant. The Mean temperature of 28°-32° C is best suited for the growth of sugarcane. Higher temperature above 45° C decreases tillering and prevents growth while temperatures below 20° C will delay development. The areas with a minimum temperature of < 5°C are not appropriate for growing sugar cane. The crop is doing well in tropical regions that receive 75-120 cm of rainfall each year. It needs a 10-18 month long, growing season. Relative humidity of 70-85 and 55-75 percent during growth stage and ripening stage is ideal respectively. Relative humidity < 50% during the growing season is not suitable for sugarcane cultivation.

Table 3. Meteorological requirements of sugarcane at different growth stages

S. No.	Growth stages	Temperature (°C)	Relative humidity (%)
(a)	Germination stage	26-32	56
(b)	Tillering stage	21-30	50
(c)	Grand growth stage	20-25	85
(d)	Differentiation stage	16-18	70
(e)	Maturity stage	8-10	65

❖❖❖

Chapter

4

CULTIVARS OF SUGARCANE

Sugarcane varieties are mainly classified into three groups –

1. **Early varieties:** The varieties attend 16.5% sucrose, and 85% purity in 10 months is kept in early category
2. **Mid-late varieties:** The varieties accumulate above 16% sucrose level and 85% purity in 12 months is grouped in a mid-late category.
3. **Late varieties:** Those varieties attend the similar 16% sucrose level and 85% purity in more than 12 months and maintain up to 14 months stage are grouped in late-maturing varieties.

Normally, two groups i.e. early and mid-late/late are observed. Farmers prefer the varieties of high tonnage and sugar mills prefer the higher sugar recovery and sucrose in cane

Early maturing varieties, lower sucrose and higher tonnage are also available nowadays and are in higher demand from sugar cane farmers as well as from sugar mill operators. These varieties perform well in the country and account for higher production of sugar

Table 4. Different varieties of sugarcane

	Varieties
Early	CoP9301, BO130, BO139, BO153,COP16437(Rajendra ganna 1), COP112, CoC671, Co7704, CoLk 94184 (Birendra), Co 6907, Co 94008, Co 86010, Co Vsi 434, Co 2001-13, Co 2001- 15, 83 A 30, Co 8013, Co 8014, 87 A 298, 86 V 96, 2003 V 46, 93 V297, 93 A 145, Co 85003 and 83 R 23, CoSe-95422(Rasbhari), CoH-2201(Haryana 92), Co-85084, Co-86249(Bhavani), Co 0403 (Samriddhi)
Mid late	BO91, COP9206, COP9302, BO136, BO137, COP 9702. BO141, BO146,COP 2061, BO154, CoP 09437, Co 86032, Co 95020, CoT1575, Co T 8201, Co A 7602, Co 7805, 85 R 186, Co M 265, Co 7219, 97 R 129, Co 99006 and Co 99006, CoS 94270, CoSe 95423 (Rajbhog), Co 8371, Co 91010 (Dhanush), CoM88121 (Krishna), Co 2001-13 (Sulabh)

Late	Co 8011, 97 R 401, Co 7706, Co 62175 and Co R 8001
Disease resistant	Co 87263 (Saryu), CoSe 95422 (Rasbhari), CoSe 95423 (Rajbhog), Co 8371, Co 94008 (Shyama), Co7508, CoA 7602, Co8014, CoR8001, 85A261, 87A298, 90A272, Co6907, 86V96, Co7706, 83V15, 89V74, 87A397, CoOr 03151 (Sabita), CoPb 08212, CoP 09437, Co99004 (Damodar), Co 2001-15, Co 0218 (Shreyas)
Insect-pest resistant	CoPb 08212,CoP 09437, BO141, Co1158, Co7218, Co7227, CoJ67, CoS767, Co975, CoC771, Co6806, Co6907, Co802,Co6304, CoS510
High ratoonability	Co 8013, Co 6907, Co 8014, 85A261, 87A298, 90A272, 92A123, 81V48, 91V83, 93V297, 97V60 and 83R23, Co 86249 (Bhavani), Co pant 90223, CoS 94270, Co 86032 (Nayana), Co 0238 (Karan 4), CoOr 03151 (Sabita)
Pollutant stress tolerant	CoC771, CoC95076
Heat and Moisture stress conditions	Co 6907, Co 95020, 83 R 23, 85 R 186, Co 99006, and Co 99004, C0449, Co617, Co1307, CoA7602, CoT8201, CoA8402, Co7717, Co8208, Co8362, Co85007, CoC6806, Co853, Co 997, Co 1103, Co 1107, CoLk 8003, Co 87263, Co 87016, BO 89, BO 90, BO 99, BO109, BO104, BO109, CoS 767 Co 98014, Co 740, Co 997, CoA 7602, MS 7054, CoM 7125, CoSi 94071, CoSi 94072, Co 94012, CoC 671, Co 91010, Co 92002, Co 92020,Co 93009, Co 97009, Co 99004, Co 8338, Co 91017, Co 200012, Co 0212, Co 0218, Co M 0265, 81A99, CoA 92081, CoA 92081, CoA 03081.
Waterlogged and Swamp conditions	Co 2061, 96 A 3, 99 A 5, 97 A 85, 84 A 125, Co 6907, Co T 8201, Co 7219, 85 A 261, 83 V 15, Co 7706, Co A 7602, 93 A 145 and Co R 8001, CoP1, Bo91, Bo99, Bo43, Co8231, Co82323, Co8145, CoT8201, Co88322, Co87268, UP 9529, UP 9530, CoS 94636, Co 03151 (Sabita), Bo 99, BO 110, BO 128, Co 8231, Co 8371, Co 98007, Co99006, Co0118, Co0232, Co0233, Co 0238, CoP 9103, CoP 9104, CoBln 9103, CoS 8118, CoTl 88322, CoSe 96436, UP 9529 and UP 9530
Saline-Alkaline soils	Co T 8201, 81 A 99, 93 A 145, Co 99006, Co 99004 and Co 7219
Low temperature stress tolerant	Co842, Co1148, CoJ46, CoJ83, CoJ79, CoJ76
Variety suitable for jaggery making	BO 70, BO 90, BO102,BO104,CoS 767, Co 1148, Co 6907, CoT 8201, CoA 7602, Co 7219, Co 7706, Co 449,Co 62175, Co 99004 and Co99006

The varieties with high fibre and moderate sugar can be used for cogeneration of electricity from bagasse. The interspecific and intergeneric hybrids of sugarcane with wild *S. spontaneum* or *E. arundinaceus* have the potential to be used as energy canes due to their ability to produce high biomass. Varieties with high cane yield and high Brix are suited for ethanol production as the total sugars are important than the sucrose content. Promising clones are expected to be evolved in the near future.

Chapter

5

FIELD MANAGEMENT

5.1. Soil requirements

It can be grown in a wide range of soil. This can grow on loams, clayey loams, black cotton soils, brown or reddish loams and even laterite soils. Sugarcane can withstand any kind of soil that could retain moisture. But deep loamy soils are great for its production and good drainage soil for sugarcane cultivation. The soil should be high in nitrogen, calcium, and phosphorus but not too acidic or alkaline. Sugarcane rapidly and extensively exhausts the soil's fertility and its cultivation requires a high dose of manures and fertilizers so fertile soil is good for the cultivation of sugarcane.

5.2. Seed bed preparation

As sugarcane crop stands in field for more than a year, it is necessary to give deep ploughing by mould board plough drawn by tractor. The proper time for ploughing is immediately after the preceding crop is harvested or just after a good shower of rain received. The land is then exposed to the atmosphere for a month. The harrowing is done 3 to 4 times to break clods and to make the land smooth and even to facilitate uniform irrigation. Four to Six ploughings to produce good tilth is recommended. Each ploughing should be followed by planking. Compost or Farm yard manure @ 15-20 tonnes per hectare should be applied in the field before 30 days of planting and mixed thoroughly with soil during land preparation. Chlorpyriphos 10 percent @12-15 kg /ha should be applied at the time of final land preparation to prevent the incidence of termites and early shoot borer, etc.

Figure 1: Seed bed preparation

5.3. Seeds and sowing

Sugarcane is propagated vegetatively by small cuttings of the stalks having one or more buds, called as seed setts. Generally, 3-budded setts are used for planting of sugarcane. Nowadays, 2-budded and 1-budded setts are also using by some farmers and scientists.

Table 5: Requirement of seeds/ha

Particulars	No. of seed setts	Seed rate
3-budded setts	35000-40000	60-80 qt/ha
2- budded setts	55000-60000	50-60 qt/ha
1- budded setts	85000-90000	25-30 qt/ha

5.4. Selection of quality seeds

Since sugarcane is cultivated by seed setts so, selection of seed setts is very important for achieving good crop stand and higher no. of milliable canes.

Criteria for selection of quality seed setts

1. Selection of high yielding variety which is resistant to insect-pest and diseases
2. The optimum length of internodes and number of buds
3. Nutritional status of the mother seed crop
4. The freshness of seed material
5. Absence of dry scales on the buds of seed setts
6. Select $\frac{1}{3}$rd to $\frac{3}{4}$th of top portion as it consists of high glucose content and richer in moisture which is helpful for quick germination.

5.5. Seed treatment

Since germination is a very important phase, governing plant population, every possible precaution must be taken to get good germination. Setts must be given proper treatment before planting for improving germination.

(a) Soaking of setts in water or a lime solution has given good results. It is, therefore, recommended that setts should either be soaked in fresh water for 24 hours and more or in a saturated solution of lime for 8 hours before planting.

(b) Treatment of setts with various hormones like Indole Acetic acid (0.5 %), Naphthol Acetic Acid @ 200ppm and acetic tricarboxylic acid has also been employed to improve the germination of setts.

(c) Some case studies reported from Padegaon, Maharastra revealed that dipping setts in 5 % sulphate of ammomia for six hours has given better germination and better yield.

(d) Sen (1954) has shown that both the germination and yield are improved when setts are soaked in 35 % molasses solution neutralized with lime overnight prior to planting. The planting setts get enriched with nutrients and carbohydrates responsible for the stimulation of the growth of sugarcane crop. By soaking setts in water, carbohydrates of setts pass to water and may interfere in growth.

Figure 2 : Sett treatment

(e) Dipping of seed setts by Bavistin @1g/L for 10-15 minutes.

(f) Dipping of sugarcane setts in neem cake slurry (1kg/5L) for 18-24 hours.

(g) Treatment of cut ends with the coaltar, termite oil, 1 % DDT solution, burnt crude oil mixed with aldrin and planted immediately.

(h) Hot water treatment of setts at 52° C for 20 minutes.

(i) Treatment of setts with 1% solution of $KMnO_4$, $MgSO_4$, or potassium ferrocyanide, 0.1-0.5 % solution of ammonium sulfate for 12 to 24 hours gives better results.

(j) Soaking of setts in cold water for 12-48 hours for improving germination.

(k) Setts should be treated by dipping them in a solution of *Tricoderma viridae* @ 5 gram per litre of water before planting to protect them from soil and seed borne diseases. 150 kg setts should be dipped in 40 litre water during chemical treatmemt.

5.6. Seed spacing

Row to row spacing in sugarcane planting is very important factor for deciding seed rate. Sugarcane setts should be planted at a distance of 90 cm with recommended dose of fertilizers. Indian Institute of Sugarcane Research suggested the closer spacing of 75 cm for cane rows without reduction in yield of sugarcane. Now a days, due to scarcity of labour, mechanization in sugarcane is promoted by different Institute, Sugar mills, traders, etc. For easy harvesting through sugarcane combine harvester, many scientists have taken 120 cm spacing in their experimental fields of sugarcane crop. Different experimental findings shows that optimum row spacing ranged from 60 cm to 120 cm under different locations and conditions. Inter-row spacing depends upon the tillering capacity of variety, planting time, soil fertility status, biotic and abiotic stress condition, etc. In general, the row spacing adopted in India is narrower than other developed countries of the world. In many sugarcane countries, it is planted at a spacing of 180 cm.

Table: 6 Recommended row spacing for sugarcane in different states of India

S.No.	States	Row spacing (cm)
1.	Uttar Pradesh	90
2.	Maharashtra	90-120
3.	Bihar	75-90
4.	Andhra Pradesh	80-90
5.	Gujarat	90
6.	Karnataka	90
7.	Tamil Nadu	80
8.	Rajasthan	75-90
9.	Madhya Pradesh	90
10.	West Bengal	90
11.	Orissa	90
12.	Haryana	60-75
13.	Punjab	60-75
14.	Kerala	90

5.7. Planting time

Sugarcane requires about 25-32° C temperature for germination. In north Indian conditions this temperature requirement is fulfilled twice, i.e. in October and February-March. Delayed planting restricts tillering because crop is bound to enter in the grand growth phase within 70-90 days after planting in response to increased atmospheric humidity and temperature with the onset of South-West monsoon by the end of June or early July. So, optimum time of sowing is an important factor in sugarcane crop.

5.7.1. Autumn planting

Sugarcane planting is generally completed in October month. Delay in planting may cause a reduction in the germination of sugarcane due to low temperatures during the winter season. Autumn planted sugarcane crop can be exploited by utilizing the spaces between two rows by planting of inter-crops such as wheat, Tonia, Fenugreek, Radish, etc.

5.7.2. Spring planting

Cane is planted in the month of February-March; sometimes planting can be delay up to April month. But delaying in planting can cause serious losses in yield due to poor crop stand. So, proper knowledge of planting time is necessary for sugarcane crop as it is an important factor.

Table 7. Planting time in different states

States	Months
Punjab and Haryana	March
Uttar Pradesh	February-March
Bihar	January –February
Tamil Nadu, Andhra Pradesh, Maharashtra and Karnataka	December-February

5.7.3. Adsali planting

In Maharashtra, Adsali planting is done during July-August months and crop duration is about 15-18 months and in Tamil Nadu, the crop is planted during July-September.

5.7.4. Eksali planting

It is common in south India. The crop is planted during January- February and harvested after one year.

5.8. Methods of planting

Planting methods differs from region to region and based on planting material used. Following planting methods are adopted in India,

5.8.1. Planting on flat beds

Generally this method is practiced in India's low rainfall zones. Using this system, shallow furrows (8-10 cm deep) are opened at a distance of 75-90 cm with a country plough or tractor powered furrow opener. After planting, the furrows are covered with 5-7 cm soil and the field is leveled by heavy planking. There must be adequate moisture in the field at the time of planting.

5.8.2. Ridge and furrow method

This method is usually adopted in areas of drainage problems. In this method, the 'V' shaped furrows are opened in about 90 cm apart and 20-25 cm deep. For this purpose tractor operated deep furrower, deep furrower cum fertilizer applicator and deep furrow sugarcane cutter planter developed by IISR, Lucknow. The setts are placed in a horizontal position in furrows in an end-to-end fashion. If the seed stalk is not healthy and internodes are longer, an eye-to-eye system of planting may be adopted. As soon as the canes start growing, the furrows are partly filled with soil and inter-row cultivation is carried out.

5.8.3. Trench method

Typically this system is used in coastal areas as well as other areas where the crop grows very tall and the heavy winds during the rainy season cause cane lodging. Trenches are dug at a distance of 75-90 cm, with the help of tractor operated trencher and trench planter or by manual labor. The trenches should be about 30 cm deep. The mixture of fertilizers (NPK) should be spread uniformly in the trenches and mixed thoroughly in the soil. The setts are planted end to end in trenches. Drenching of setts with chlorpyriphos 20 EC (1 kg a.i./ha or 5 liters/ha) to protect from the soil-borne insects is required. The trenches are filled up with loosened soil after planting.

5.8.4. Furrow Irrigated Raised Bed system (FIRB)

In western India, most prevalent cropping system is under wheat-sugarcane- ratoon-ratoon-wheat. Accordingly sugarcane is planted in the last weeks of February- first weeks of March, sometimes upto April after harvest of Wheat. A drastic reduction in the yield of sugarcane is a major issue when planted late after wheat harvest. In order to improve productivity of sugarcane under such cropping system, the Furrow Irrigated Raised Bed (FIRB) system has been developed by IISR, Lucknow. In this system, three rows of wheat in November are sown on raised beds. Sugarcane is planted at 80-85 cm apart furrows in February in a furrow. Sugarcane gives about 30% higher cane yield as compared to the wheat-sugarcane sequential system, without a reduction in wheat yield. In this system, irrigation is applied only in furrows, requires a lower volume of water, and provides about 20% water saving and the system increases water use efficiency by saving irrigation water. No extra irrigation is needed for field preparation for sugarcane planting. In this system, weed problem is less in raised beds and weed infestation mostly occurs in furrows which can be controlled easily by hand weeding or herbicidal spray

Furrow irrigated raised bed method

Ring pit method

Paired row planting

Poly bag technology

Bud chip method

Bud chipping machine

Figure 3 : Different planting methods of sugarcane

5.8.5. Ring Pit planting

Several progressive farmers also do the pit planting. In this method, pit dimensions of 2.25 feet in diameter and 1.25 deep depth and seed rate of 21, two budded set per pit were found to be the most economical with cane yield of 1.5 to 2.0 times higher than traditional planting process. For higher germination and cane yield in pit planting i.e. placing the setts in pits followed by light irrigation in the individual pit and then covering with 2-3 inch loose dry soil when the water disappears in pit planting is the best practice. New planting technique i.e. wide bed & furrow method (trench method) found suitable for mechanization, saving of water and seed in sugarcane with comparable yields.

5.8.6. Spaced Transplanting (STP) Technique

In the STP method, raised nursery beds of 3 feet width and convenient length are prepared. FYM is applied to a nursery bed and mixed with soil. An area of 100 sq. m is required to raise nursery for planting one hectare sugarcane. Disease-free setts are selected from nursery crops. Single bud setts are prepared, dipped in 0.2% Carbendazim for 5 minutes and then planted vertically on the raised bed. Apply Chlorpyriphos or Imidacloprid (1 ml/liter of water) on nursery beds. The setts are covered with loose soil and instantly irrigated or sprayed using Rose can. To obtain better germination and weed-free nursery, daily weeding and watering are necessary. After 6-7 weeks, the settlings are uprooted.

5.8.7. Bud Chip technique

The nursery is prepared using bud chips in this process and plants are transplanted in the main field. From seed nursery, pick disease-free canes. Use the bud chipping machine to take the bud out along with a portion of the nodal portion. A bud chipping machine was designed and developed by SBI Coimbatore. For planting in one hectare only 15 quintals of bud chips are necessary. The remaining portion of the stalk can be sent for milling. Treat the bud chips in chlorpyriphos (2 ml/liter of water) + Carbendazim (2 g/liter of water) for 5 minutes and incubate in a moist gunny bag for overnight (priming). Prepare a homogenous mixture with equal quantity of soil, sand and FYM, fill in perforated polythene bags or cavity trays, plant the bud chips in an upright position facing the bud up, cover it with soil mixture. Provide frequent watering with Rose Can and spray 1 percent urea solution after 15th and 25th day of planting. The seedlings are six to eight weeks fit for transplantation. Settling in the field will have a high success rate because it is transplanted along with earth ball. Bud chips are easy for long-distance transport. Seed cane is saved (quantity and cost) and time is 6-8 weeks. The higher rate of survival after transplantation, uniform tillering and higher yields were achieved by bud chip planting.

5.8.8. Skip furrow planting

This method of planting was adopted in Orissa. It is an amalgamation of flat and trench method. Trenches of 45-30 cm apart are dug and a gap of 90 cm is left after every two rows of cane. Earthing is done by taking two rows together, leaving an irrigation channel after every two rows of cane.

5.8.9. Sablang or sprouting method

This method is practiced in Java and Cuba. Plants grown in fertile soil with wide spacing give rise to roots from tiller are separated from the mother plant and planted separately. After this, the mother plants continue tillering and new tillers are also separated.

5.8.10. Rayungan method

After 30–40 days of planting, seed stalks are topped off. Lateral shoots grow and separate from the mother plant, and are planted in 30 cm deep and 90 cm apart trenches filled with manure by mixing soil. IISR, Lucknow, had developed this system.

5.8.11. Tjeblock method

Tjeblock method is an improved method of rayungan. At its half-length, the stalk is cut off and planted vertically with one node under the soil for planting, and properly irrigated and fertilized. After some time the upper buds of both Tjeblocks and mother cane sprout and are planted as rayungans by cutting them into pieces. The best consideration for the best supply of energy and nutrient to all buds has been taken in this process.

5.8.12. Distant planting method

In this method the top setts are collected and put in nursery and after they have sprouted and rooted out. They are transplanted in the final field at a distance of 90 cm x 50 cm.

5.8.13. Algin method of sugarcane planting

Last nodes of canes are collected during cane stripping, then planted in rows to the wheat field after every four rows of wheat at a distance of 90 cm x 50 cm. Then irrigate and fertilize the grounds. After wheat harvesting, the sugarcane crop is taken care of and intercropped with urd or moong bean during summer. Scientists at the Allahabad Institute of Agriculture have evolved this method.

5.8.14. Partha method

This technique was developed by Mr. S.V. Parthasarathy an eminent sugarcane scientist. He suggested for waterlogged or excess soil moisture conditions (coastal Andhra Pradesh and Tamil Nadu during N-E monsoon period). In this process, in a slanting position, 60° to the vertical, in the wet furrow, or halfway on the ridges, are planted three eyes budded setts. Typically one eye bud is forced into the soil, and the other two will be above, and will sprout. If the moonsoon recedes, the sprouted in situ setts are pushed down into the soil and allowed horizontally to stand. Soil is spread at the surface. The crop is manured at this stage.

5.8.15. Cane node technology

In this method, cut the healthy viable nodes along with root bands about 4-5 cm in length from sugarcane plants. Dip the cane segments in 0.10-0.75 % liquid solution of carbendazim for 15 minutes. Prepare liquid organic formulation made up of cattle dung and urine in a ratio of 1:10 and dilute it 5-10 times with water for priming and stirred to make slurry. Planting material is immersed in the slurry for 15 minutes. Then prepare soil mixture of field soil and decomposed FYM in the ratio of 1:1 to 1:5 and maintain moisture content of 50-60%. Spraying of chlorpyriphos (1-10ml/L) on the soil mixture should be done. For nursery, area should be 5.0 x 0.60 m^2 is sufficient for one hectare area. Primed cane nodes should be kept in incubation for 5-7 days so that buds will start swelling and sprouts after 5-7 days. Sprouted buds should be planted in adequate spacing with agronomical practices.

The benefit of this technology is the saving of seed material as it requires only space should be there 10 -12 quintals as compared to 70-80 quintals per hectare in conventional practices. Seed cane multiplication ratio (1:40) is higher in cane node technology. This technology facilitates quick germination by planting sprouted cane nodes in the field.

5.8.16. Polybag technology

Mixture of field soil, cowdung / FYM, sand should be prepared in a ratio of 1:1:1 and treat with insecticide Chlorpyriphos 20 EC@ 8-10 ml/quintal and then, filled in polythene bags. Healthy undamaged bud should be cut from freshly harvested sugarcane with the help of eye bud cutter. About 24000 single bud pieces are required for raising nursery for transplanting one hectare area. Bud should be soaked in Bavistin (0.2 %) solution in 50 litres water for 5 minutes to prevent the buds from fungal infections and for enhancement of germination, development of roots and shoots add growth regulators such as Gibbrellic acid (GA_3) and Naphthalic acetic acid (NAA). After bud treatment, the bud pieces should be placed in polythene bags containing soil mixture in vertical or horizontal position. A light irrigation should be given to the plots so that the water absorbed by soil mixture. Cover the nursery with sugarcane /banana leaves to maintain the humidity. Sprinkled the water at 5-7 days interval in polythene bags till germination.

Chapter

6

CROP DIVERSIFICATION

The cropping patterns used on a farm and their interaction with farm resources, other farm enterprises and available technology determine their makeup.

The common cropping systems followed in north India are given below.

1. Rice - sugarcane - wheat
2. Cotton - sugarcane - ratoon - ratoon
3. Rice - wheat - rice - sugarcane
4. Rice - mustard - sugarcane - ratoon
5. Maize - wheat - sugarcane
6. Rice - chick pea - sugarcane - ratoon
7. Rice - sugarcane - ratoon
8. Cotton - wheat - *Rabi* sorghum - sugarcane

In southern India, sugarcane based common cropping systems is as follows.

1. Rice - sugarcane - ratoon - wheat
2. Cotton - sugarcane
3. Cotton - sugarcane - ratoon
4. Rice - groundnut - sorghum - ragi - sugarcane
5. Sugarcane - ratoon - paddy
6. Cotton - sugarcane - chick pea
7. Rice - sugarcane - chick pea
8. Rice - potato - spring sugarcane - moong bean

6.1. Types of cropping systems

6.1.1. Mono cropping

A system that are growing same crop year after year on the same piece of land. Cropping intensity in India is always 100%. Monocropping allows for more insect-pest disease incidence.

6.1.2. Multiple cropping

Growing of two or more crops is on the same piece of land in a year. It is based on intensification of cropping in time and space dimensions i.e. more crops within a year and more no. of crops on the same piece of land at any given period.

Multiple cropping is of different types

(a) Sequential cropping

(b) Overlapping/ relay cropping

(c) Intercropping

(d) Paira/ utera cropping

(e) Mixed cropping

(f) Parallel cropping

(g) Multistorey cropping/Multitier cropping

6.2. Sugarcane intercropping in sustainable production

6.2.1. Intercropping with autumn planted sugarcane

Sugarcane + potato

Sugarcane + carrot

Sugarcane + mustard

Sugarcane + chilli

Sugarcane + radish

Sugarcane + french bean

Sugarcane + cauliflower

Sugarcane + sugarbeet

Sugarcane + wheat

Sugarcane + coriander

Sugarcane + niger

Sugarcane + maize

6.2.2. Intercropping with spring-planted sugarcane

Sugarcane + moong bean

Sugarcane + urd bean

Sugarcane + cowpea

Sugarcane + sunnhemp

Sugarcane + cucurbits

Sugarcane + okra

Sugarcane + brinjal

Sugarcane + French bean

Sugarcane + Sugarbeet

Sugarcane + Tomato

Sugarcane + Maize

Sugarcane + Wheat

Sugarcane + Bengal gram

Figure 4: Sugarcane based intercropping system

Weeds positive allelopathy, higher net return due to reduced cost of cultivation, additional nitrogen fixation through leguminous intercrops. However, there are some disadvantages of intercropping. These are as follows.

1. Mechanization becomes difficult in intercropping system

2. Pink bollworm insect of wheat crop can migrate to the sugarcane crop if wheat is taken as Intercrop to some exent.

3. Nutritional deficiency in plants may persist if additional nutrient fertilizer is not applied for intercrops.

Packages and Practices of Different Intercrops Sown with Sugarcane.

1. Sugarcane + garlic: Intercropping of sugarcane with garlic has been found very beneficial for fetching higher income. In this intercropping system, sugarcane is planted at 90 centimeter distance in the last week of September or first week of October with the application of manures and fertilizers in the furrows and two rows of garlic between the rows of sugarcane at a distance of 30 cm. An extra dose of nitrogen should be applied to the garlic crop. Irrigation should be given as per need of crop. The garlic crop can be harvested in the month of January/ February.

2. Sugarcane + coriander: Sugarcane is planted in rows 90 centimeter apart, in the month of November. Recommended dose of fertilizer should be given at the time of planting. Two rows of coriander at 30 centimeter apart in the centre of sugarcane rows leaving 30 centimeter distance between cane and coriander rows are sown. No extra fertilizer should be applied for coriander. Picking of coriander at Green leaf stage can be done which can fetch additional income to the farmers and it can be also harvested for seed purpose at later stage.

3. Sugarcane + green gram: Two rows of short duration moong is planted in between the rows of sugarcane in the month of march at 30 centimeter apart. As moong is legume/pulse crop, so there is no need of extra dose of fertilizer. Harvesting of moongbean/green gram can be done in the month of May-June. This intercropping helps in increasing nitrogen content in the soil.

Chapter

7

NUTRITION MANAGEMENT IN SUGARCANE

Sugarcane is a longest duration crop which takes about 10-12 months for their growth and development. Sugarcane is a heavy feeder of nutrients. A crop producing 80 tonnes per hectare removes approximately 90 to 110 kg nitrogen (N), 180 to 300 kg of phosphoric acid, 60 to 150 kg potassium from the soil. Thus, adequate and timely manuring of sugarcane crops is necessary to secure a high yield of good quality cane. The use of manures and fertilizers plays an important part in the maintenance of soil fertility and a good stand of the sugarcane crop. The fertilizer and manure requirement of sugarcane depends upon the soil and climatic conditions. Quantity of nutrients should be based on a soil test. There is a wide range of fertilizer and manures which can be used for achieving higher sugarcane yields and productivity per unit area. These are as follows.

7.1. Organic manure

7.1.1. Bulky organic manures:

Bulky organic manures are those manures which contain less percentage of nutrients and are required in higher quantity. Examples are as follows: Farmyard manure (FYM), Compost (Farm compost, Town compost, Azo-compost, Pressmud compost).

7.1.1.1. Farmyard manure

Farmyard manure is manure made up of waste materials, usually dung and urine, created by farm animals, most often by cows and buffaloes. The waste products contain a wealth of nutrients, including the crucial element that plants need nitrogen, phosphorus, potassium, micronutrients. Three to four weeks before sowing crops, partially decomposed manure from the farmyard can be applied to the soil via manure spreader or by the broadcasting method. This access to the elements helps the manure to decompose further and provide nutrients for the soil. The nutrients may be washed if there is a long pause between placing the manure on the soil and sowing the crops. FYM contains on an average about 0.5 percent Nitrogen, 0.15 % Phosphorus and 0.5 % Potash under Indian conditions. In sugarcane, FYM should be applied about 15-20 tonnes per hectare at one month before planting.

There are several factors which influenced the composition of FYM are as follows.

1. Source of manure
2. Age of the livestock
3. Food and nutrition of animals
4. Function of the animals
5. Storage of farmyard manure
6. Nature of litter

Benefits of farmyard manure

- Farm Yard Manure is a powerful soil improver that strengthens and retains a variety of soil's natural properties;
- Enhances soil fertility
- A safe supply of nitrogen now available
- Adds humus and slow nutrient release into the soil
- Aids in water and conservation of nutrients
- Helps break down of heavy fields
- Adds structure to light and shallow soils
- Attracts worms into the earth
- Good for mulching

7.1.1.2. Compost

The word compost literally means a substitute of natural product. Composting is the method of decomposing vegetable and animal waste (rural or urban) and during a quickly utilizable state for increasing and preserving soil fertility and decomposing material is called compost. Compost should be applied prior to 15 days of sugarcane planting at the rate of 7-15 tonnes per hectare. Compost is of different kinds.

1. Farm compost/ Rural compost contains about 0.5 % N, 0.15 % P_2O_5 and 0.5 % K_2O
2. Town compost is rich in nutrients (1.4 % N, 1.0 % P_2O_5 and 1.4 % K_2O
3. Azo-compost

4. Water hyacinth compost
5. Press mud compost
6. Mushroom spent compost (SMC) 1-2 % N, 0.20 % P_2O_5 and 1.4 % K_2O

7.1.1.3. Vermicompost

Vermicompost is a result of the decomposition process using various worm species, typically red wigglers, white worms, and other earthworms (epigeic and endogeic) to produce a mixture of decomposing vegetable or food waste, bedding products, and vermicast. This method is termed vermicomposting, and to this end, the rearing of worms is termed vermiculture. Vermicompost contains water-soluble nutrients, and is an ideal organic fertilizer and nutrient-rich soil conditioner. This is used in sustainable organic agriculture and small-scale farming.

Advantages

1. Improves aeration of the soil
2. Enriches the soil with microorganisms (adding phosphatase and cellulase enzymes)
3. Microbial activity in worm castings is 10-20 times higher than in soil and organic matter ingested by the worm.
4. Attracts deep-burrowing earthworms in the soil
5. Improves the storage efficiency of water into the soil.

Table 8: Nutrient content in vermicompost

Nutrient	Percentage (%)
Organic carbon	15-25
Nitrogen	1.3-2.6
Phosphorus	0.7-1.8
Potassium	0.4-0.8
Micronutrient	mg/kg
Zinc	2.0-13
Manganese	3.8-19.5
Iron	2.5-21.5
Copper	1.9-6.0

7.1.14. Green manure

Green plant material of the leguminosae/ nonleguminosae family used for manure is considered green manure. This provides the soil nutrients and improves soil health, in addition to providing organic matter.

Benefits:

1. Manures improve physical properties in the soil.
2. This provides mineral nutrients (macro and micro), and increases the supply of nutrient.
3. It acts as a storehouse of food for soil micro-organisms.
4. It helps in the prevention of nutrient loss and soil erosion.

It is of two types

1. Green manuring

Growing of plants in the fields and ploughed-in after few days for decomposing of plant material is called as green manuring. Green manure crops fix nitrogen and increases nutrient content especially Nitrogen, Phosphorus and little amount of micronutrients. For green manuring, pulse crops are generally used. Pulse crops in rotation can be an effective way of improving soil fertility and productivity for a long term basis. Some examples of green manure crops are as follows. *Sesbania aculeata* (Dhaincha), *Crotolaria juncea* (Sunnhemp), *Sesbania rostrata* (root and stem nodulated legume), *Vigna sinensis* (Cowpea), *Melilotus alba* (Senji), *Cymopsis tetragonoloba* (Cluster bean), *Vigna radiata* (Moong bean), *Vigna mungo* (Urd dal), *Arachis hypogea* (Groundnut or Peanut), *Lens culinaris* (Lentil), *Pisum sativum* L. (Pea), *Cajanus Cajan* L. (Tur / Arhar/ Pigeon pea etc.

Table 9: Amount of nitrogen fixed (kg/ha) by green manure crop

Crop	Age	Nitrogen fixed (kg/ha)
Sesbania aculeata (Dhaincha)	60	77.7
Crotolaria juncea (Sunnhemp)	60	84.4
Sesbania rostrata (root and stem nodulated legume)	50	96.0
Vigna sinensis (Cowpea)	60	74.0
Melilotus alba (Senji)	60	134.5
Cymopsis tetragonoloba (Cluster bean)	50	91.0

2. Green leaf manuring

Green leaf manures (GLMs) are organic manures made from leaves gathered from elsewhere. The use of GLMs crops is not a recent technique. Green leaf manure can be described as the pruning and gathering of green leaves, twigs, stems from trees, shrubs and herbs and then being applied as fertilizers. Forest tree leaves are major sources of these manures while some other sources are herbs and a shrub growing in the boundary of field as fences and wastelands etc. uses of Greenleaf manuring is common practices in hills and terrains.

Examples: Gliricidia *(Gliricidia maculata),* Karanj *(Pongamia pinnata),* Mahua *(Madhuca indica),* Acacia *(Acacia Senegal),* Subabool *(Leucaena leucocephala),* Cassia *(Cassia auriculata),* Gulmohar *(Delonix regia),* AaK *(Calotopis gigantia), etc.*

Table 10. Nutrient Percentage in Green Manure Crops and Green leaf Manure

	Nitrogen (N) %	Phosphorus (P_2O_5) %	Potassium (K_2O) %
Green Manure Crops			
Sunhemp *(Crotalaria juncea)*	2.30	0.50	1.80
Dhaincha *(Sesbania aculeate)*	3.50	0.60	1.20
Sesbania *(Sesbania speciosa)*	2.71	0.53	2.21
Cowpea *(Vigna sinensis)*	0.70	0.10-0.20	0.60
Guar *(Cymopsis tetragonoloba)*	0.30	-	-
Horse gram (*Mactotyloma uniflorum*)	0.33	-	-
Mothbean	0.80	-	-
Green gram (*Vigna radiata*)	0.72	0.18	0.53
Sunnhemp (*Crotolaria juncea*)	0.80	0.10	0.50
Black gram (*Vigna mungo*)	0.85	0.18	0.53
Green leaf Manure			
Gliricidia (*Gliricidia sepium*)	2.76	0.28	4.60
Pongania (*Pongamia glabra*)	3.31	0.44	2.39
Neem (*Azadirachta indica*)	2.83	0.28	0.35
Gulmohur (*Delonix regia*)	2.76	0.46	0.50
Peltophorum (*Peltophorum ferrugenum*)	2.63	0.37	0.50
Calotropis (*Calotropis gigantea*)	2.06	0.54	0.31
Cassia (*Cassia fistula*)	1.60	0.24	1.20

Source: Fertilizer statistics, 2003-04, and Handbook of Fertilizer Usages, FAI, New Delhi

It is important to note that about 8-10 weeks after planting green manure crops should be incorporated in the soil and approximately 50-60 kg Nitrogen is fixed. In the case of spring-planted sugarcane moong bean and autumn planted sugarcane rabi season pulse crops is taking as an intercrop which can be utilized as green manure crops after 2-3 picking of pods.

7.1.2. Concentrated organic manures/ organic nitrogen fertilizers

Concentrated organic manures contain a higher percentage of nutrients than bulky organic manures. Examples of concentrated organic manures are oilcake and bonemeal (raw and steamed), blood meal, etc.

7.1.2.1. Oil cake

Oilseeds are a good source of oilseed cakes. After extraction of oil, the solid portion is left and called oilseed cakes can be used as manures. Oilseed cakes can be classified into two types. (Table 11)

1. Edible oilseed cakes can be used for feeding of livestock
2. Non-edible oilseed cakes can't be used for feedstock of livestock

Table 11: Nutrient content in oilseed cakes

Oilseed cakes	Nutrient content (%)		
	Nitrogen (N)	Phosphorus (P_2O_5)	Potassium (K_2O)
Edible oil cakes			
Groundnut cake	7.3-7.5	1.3-1.5	1.3-1.5
Sesame seed cake	5.5-6.0	2.0-2.1	1.2-1.3
Niger seed cake	4.7-5.0	1.8-2.0	1.3-1.5
Linseed cake	4.9-5.0	1.4-1.6	1.3-1.5
Safflower cake (Decorticated)	7.9-8.0	2.2-2.5	1.9-2.0
Coconut cake	3.0-3.2	1.9-2.0	1.8-2.0
Cotton seed cake (Decorticated)	6.4-6.5	2.9-3.0	2.2-2.5
Mustard seed cake	5.2-5.5	1.8-2.0	1.2-1.5
Non-edible oil cakes			
Castor cake	4.3-4.5	1.8-2.2	1.3-1.5
Karanj cake	3.9-4.0	0.9-1.0	1.2-1.5
Neem cake	5.2-5.5	1.0-1.2	1.4-1.6
Mahua cake	2.5-2.8	0.8-1.0	1.8-2.0
Safflower cake (undecorticated)	4.9-5.0	1.4-1.6	1.2-1.4
Cotton seed cake (undecorticated)	3.9-4.0	1.8-2.0	1.6-1.8

Source: Fertilizer statistics, 2003-04, and Handbook of Fertilizer Usages, FAI, New Delhi

On the manurial response of sugarcane, much data had accumulated in the different growing states of India.

7.2. Fertilizers

Fertilizer is any natural or synthetic substance (other than liming materials) that is added to soil or plant tissue in order to provide one or more plant nutrients necessary for plant production. There are several sources of fertilizer, developed both naturally and industrially.

7.2.1. Primary nutrient fertilizer

7.2.1.1. Nitrogenous fertilizers

Nitrogen is a significant structural part of the cell and also of the cell wall (5 percent N) and plays an important role in plant metabolism as an integral component of a number of metabolically active compounds (e.g., amino acids, proteins, nucleic acids). Nitrogen is a nutrient for sugarcane crops that induces growth. It is a fundamental factor for determining yield and increasing water use and radiation use efficiency. This ingredient is useful in fertilizers, particularly during the middle stages of the lifetime of a plant, when it requires encouragement to start growing large leaves and stem new ones. Plant absorbs N from the soil solution as nitrate (NO_3^-) and ammonia (NH_4^+), e.g., Rice, Tea. These ions are released in the soil upon decomposition and mineralization of organic matter (OM) and applied manures by the micro-organisms and directly from the applied N-fertilizer. Both organic and inorganic fertilizers have nitrogen sources in them.

(1) **Amide fertilizers:** These fertilizers are carbon compounds and called as organic fertilizers. Amide fertilizers contain nitrogen in the form of amide. Important fertilizers are as follows: Urea (46 %N), calcium cyanamide (21-22 % N).

(2) **Ammonia containing fertilizer :** Fertilizers in which Nitrogen is present in the form of ammonium ion. Examples: Anhydrous ammonia (82% N), Ammonium chloride (26% N), Ammonium sulfate (20.6% N), Ammonium phosphate (20 % N), Aqueous ammonia (20-28 % N)

(3) **Nitrate containing fertilizer :** Fertilizers in which Nitrogen is present in the form of nitrate. Examples: Ammonium nitrate, Calcium nitrate (15.5 % N), Sodium nitrate (16 % N)

(4) **Both Ammonia and Nitrate containing fertilizer :** Ammonium sulfate nitrate (24-26% N), Calcium ammonium nitrate (25%), Ammonium nitrate (32-35 % N), Ammonium sulfate nitrate (24-26 %),

About 185 Kg nitrogen is recommended to the sugarcane cultivation. Half nitrogen should be applied at the time of sowing and rest ¼th at 30 days after sowing and ¼th at tillering stage

(5) **Slow release nitrogen fertilizer (SRNF):** It is a granulated fertilizer that releases nutrients gradually into the soil (i.e. with controlled release period). The slowness of the release is determined by the low solubility of the chemical compounds in the soil moisture. SRNF are typically coated or encapsulated with inorganic or organic materials that control the rate, pattern, and duration of plant nutrient release. Most importantly, the release rate of a slow release nitrogen fertilizer is designed in a pattern synchronized to meet the changing crop nutrient requirements.

(6) **Nitrification inhibitors (NI):** Nitrification inhibitors delay / inhibit the ammonium ion bacterial oxidation (NH_4^+) by depressing the activity of Nitrosomonas bacteria in the soil over a certain period of time (4-10 weeks). The aim of using nitrification inhibitors is to control the loss of nitrate by leaching or by denitrification of nitrous oxide (N_2O) from the top of the soil by keeping N in the long ammonium form longer and thus increasing the NUE. In addition, by delaying the conversion of ammonium to nitrate, nitrification inhibitors avoid undesirable high levels of nitrate in plants used for human and animal nutrition. Nitrification inhibitors favour plant partial ammonium nutrition because plants require less energy to incorporate ammonium into amino acids; nitrate must first be reduced to ammonium, which requires energy.

Nitrification inhibitors are based on product's ability to tie-up copper- a critical metal used by nitrification bacteria.

7.2.1.2. Phosphatic Fertilizers

Phosphorous is a second essential nutrient. Soil is the principal plant nutrient source of P. Plant roots absorb P mainly as orthophosphate di-hydrogen ($H_2PO_4^-$), and under neutral to alkaline conditions as orthophosphate (HPO_4^{2-}) ions from the soil solution. Phosphorous is a nutrient that is desperately required for sugarcane plants to stabilise the root and shoot system throughout its lifecycle. Phosphorous deficient sugarcane plants may undergo stunted growth. Phosphorous is long-lasting and slow acting. Using fertilizer in the field before planting of sugarcane is generally a good idea.

(1) **Water soluble phosphoric acid fertilizer:** Fertilizers which are soluble in water and easily available to the plants.
Examples: Single super phosphate (16-18 % P_2O_5), Double super phosphate (32 % P_2O_5), Triple Super phosphate (48 % P_2O_5), Ammonium phosphate (20 % P_2O_5)

(2) **Citric Acid soluble phosphoric acid fertilizer:** Fertilizers which are soluble in acid and basic in reaction particularly suitable for lateritic and acid soils because in such soil phosphoric acid converts into the mono-calcium phosphate. These fertilizers are generally used in long duration crop like sugarcane.
Examples: Basic slag (3-8 % P_2O_5), Di-calcium phosphate (34-38 % P_2O_5), Rhemania phosphate (23-26 % P_2O_5)

(3) **Water and acid insoluble**
Rock phosphate (20-30% P_2O_5)

Phosphorus should be applied at the time of sowing about 80 kg P_2O_5/ ha in sugarcane from different sources of phosphorus.

7.2.1.3. Potassic Fertilizers

Potassium allows the sugarcane plants to develop stronger with healthier roots. It can also help to reduce injury to the plants as other nutrients are exhausted. This resource is important for photosynthesis in sugarcane leaves, and it will slow down any pathogens that may infect the plant. The Potassium fertiliser has many benefits. Potassium is known to increase the quality of many crops such as sugarcane in addition to providing resistance to environmental stress. Plants absorb potassium as K^+ ion in larger quantities from the soil. Plants can store K^+ in huge amounts and accumulate in their tissues than what is needed without causing toxicity which is called as **luxury consumption**. When and how this fertilizer should be used depends on what you are planting. If there is a potassium deficiency in plant, you may see yellowing or browning on the edges of leaves. Leaves will eventually die off if the deficiency continues.

These potassic fertilizers are as follows

(a) **Fertilizer having K in the chloride form:** Potassium chloride / muriate of potash (KCl) containing about 60-63 % K_2O.It is cheaper than other potassic fertilizers and imported from other countries and extensively used in all crops except sugarcane, sugarbeet. In sugar crops, accumulation of sugar is affected due to chloride ion present in the fertilizers. So, use of MOP should be avoided in sugarcane crop.

(b) **Fertilizer having K in non-chloride form:** These are given below. Potassium sulphate (48-52 % K_2O), Potassium magnesium sulphate (22-24 % K_2O), Potassium nitrate or salt petre or nitre (44 % K_2O), potassium meta phosphate (40 % N). It is highly recommended to the sugarcane crop cultivation.

7.2.2. Secondary nutrient fertilizers

No special fertilizers are manufactured simply to supply Calcium, Magnesium and Sulphur, but many fertilizers contain these nutrient elements in sufficient quantities. These are also essential for plant growth and development. Sulphur is second deficient nutrient after nitrogen. Secondary nutrient fertilizers are as follows.

7.2.2.1. Calcium containing fertilizers

Calcium fertilizers are indirectly used in the soil through fertilizers and soil amendments and directly in the form of lime. Calcium is taken by plants in the Ca^{2+} forms. Some fertilizers which are used in sugarcane are as follow: Calcium nitrate (27 % Ca), Super phosphate (27-46 % Ca), Gypsum (40 % CaO), etc.

7.2.2.2. Magnesium containing fertilizers

Magnesium is generally obtained from dolomite, a material used to correct acidity. It is taken by plants in the form of Mg^{2+} ions. Few fertilizers containing Magnesium are as follow. Epsom salt (9.6 % MgO), Potassium magnesium sulphate (11.1 % MgO), Dolomite (5-20 % MgO), Magnesium sulphate (16 % MgO), etc.

7.2.2.3. Sulfur-containing fertilizers

Sulphur is required in same quantities of phosphorus. Plant is absorbed suplhur in the sulphate (SO_4^{-2}) ion forms. Important fertilizing materials supply sulphur is as follows: Ammonium sulphate (26 %S), Potassium sulphate (18% S), etc.

Figure 5: Fertilizer application in sugarcane fields

7.2.3. Micronutrient Fertilizers

Molybdenum, zinc, boron, and copper etc., are the principal micronutrients. Such elements are supplied as salts and are water soluble. Iron poses unique challenges as it absorbs in high soil *p*H and phosphate concentrations into insoluble (bio-inaccessible) compounds. Therefore, iron is also treated as a chelate complex, e.g. the derivative of EDTA. Basal application of zinc sulphate at the rate of 37.5 kg / ha. About 5 percent foliar sprays of zinc sulphate with 1 percent urea at 15 days interval till symptoms of deficiency disappear. Basal application of ferrous sulphate of 100 kg / ha. If deficiency symptoms appear in sugarcane 1% ferrous sulphate foliar spray with 1% urea at 15 days interval till deficiency symptoms disappear. Soil application of $CuSO_4$ @ 5 kg/ha in copper deficient soils. Alternatively foliar spray of 0.2% $CuSO_4$ twice during early stage of crop growth by soil application of 25-50 kg/ha $FeSO_4$ or by foliar spray of 2% $FeSO_4$.

Table 12. Nutrient percentage in macro nutrient fertilizers

S. No.	Fertilizers	Nitrogen	Phosphorus	Potassium	Sulfur	Calcium/ Magne-sium
1.	**Primary nutrient fertilizer**					
(A)	**Nirogenous Fertilizer**					
1.	Urea	46.0	-	-	-	-
2.	Ammonium sulfate	20.6	-	-	23.0	-
3.	Ammonium chloride	23.0-25.0	-	-	-	-
4.	Calcium ammonium nitrate	25.0	-	-	-	-
5.	Ammonium sulfate nitrate	23.0	-	-	-	-
6.	Anhydrous ammonia	85.0	-	-	-	-
(B)	**Phosphatic Fertilizer**					
1.	Single super phosphate (SSP)	-	16.0	-	12.0	18.5
2.	Double super phosphate (DSP)	-	32.0	-	-	-
3.	Triple super phosphate (TSP)	-	48.0	-	-	-
4.	Basic slag	-	3.0-8.0	-	-	-
5.	Rock phosphate	-	20.0-30.0	-	-	-
(C)	**Potassium Fertilizer**					
1.	Muriate of Potash (MOP)	-	-	60.0	-	-
2.	Sulphate of Potash (SOP)	-	-	48.0-52.0	-	-
(D)	**Complex Fertilizer**					
1.	Di-ammonium phosphate (DAP)	18.0	46.0	-	-	-

2.	Urea ammonium phosphate (UAP)	28.0	28.0	-	-	-
3.	Ammonium phosphate	16.0	20.0		-	-
4.	Nitro-phosphate potash	15.0	15.0	15.0	-	-
5.	IFFCO	10.0	26.0	26.0	-	-
2.	**Secondary nutrient fertilizer**					
(A)	**Calcium containing Fertilizer**					
1.	Calcium ammonium nitrate	25	-	-	-	-
2.	Dolomite	-	-	-	-	20-45
3.	Agricultural limestone	-	-	-	-	80-95
4.	Gypsum	-	-	-	-	40
5.	Single Super Phosphate	-	-	-	-	25-30
(B)	**Magnesium containing Fertilizer**					
1.	Dolomite	-	-	-	-	5-20
2.	Chelated Mg	-	-	-	-	2-10
3.	Magnesium sulphate	-	-	-	-	16
4.	Magnesite	-	-	-	-	40
(C)	**Sulphur containing Fertilizer**					
1.	Elemental sulphur	-	-	-	98.7	-
2.	Single Super Phosphate	-	-	-	12	-
3.	Gypsum	-	-	-	13-18	-
4.	Iron pyrites	-	-	-	22-24	-
5.	Ammonium sulphate	-	-	-	15	-
6.	Potassium sulphate	-	-	-	18	-

Table 13. Nutrient percentage in micronutrient fertilizer

S. No.	Micronutrient Fertilizer	Fe	Cu	Mn	Zn	N	B	Mo
(A)	**Iron containing fertilizers**							
1.	Ferrous sulphate	19.0	-	-	-	-	-	-
2.	Ferrous ammonium sulphate	14.0	-	-	-	-	-	-
3.	Fe-EDTA	12.0	-	-	-	-	-	-
4.	Fe-DTPA	10.0	-	-	-	-	-	-
(B)	**Zinc containing fertilizers**							
1.	Zinc sulphate	-	-	-	21.0	-	-	-
2.	Zinc EDTA	-	-	-	12.0	-	-	-

3.	Zincated urea	-	-	-	2.0	43	-	-
4.	Zinc oxide	-	-	-	78.0	-	-	-
(C)	**Copper containing fertilizers**							
1.	Cuprous oxide	-	89	-	-	-	-	-
2.	Cupric oxide	-	75	-	-	-	-	-
3.	Cu-EDTA	-	13	-	-	-	-	-
(D)	**Manganese containing fertilizers**							
1.	Manganese sulphate	-	-	30.5	-	-	-	-
2.	Manganese oxide	-	-	62.0	-	-	-	-
3.	Manganese chloride	-	-	7.0	-	-	-	-
4.	Mn-EDTA	-	-	12.0	-	-	-	-
5.	Mn frits	-	-	12.0	-	-	-	-
(E)	**Boron containing fertilizers**							
1.	Borax	-	-	-	-	-	10.5	-
2.	Solubor	-	-	-	-	-	19.0	-
3.	Boronated single superphosphate	-	-	-	-	-	0.18	-
4.	Boric acid	-	-	-	-	-	17.0	-
5.	B frits	-	-	-	-	-	2-6	-
(F)	**Molybdenum containing fertilizers**							
1.	Ammonium molybdate	-	-	-	-	-	-	52
2.	Sodium molybdate	-	-	-	-	-	-	39.5

Soil application of 0.5-3.0 kg B per hectare and Foliar sprays of 0.2-0.3 percent solution of solubor or boric acid should be done for boron deficiency. Soil application of 75-100 g per hectare, foliar sprays of 0.1-0.3 percent or seed treatment (50-100 g per hectare as liquid or slurry is found effective to prevent Molybdenum deficiency.

7.3. Bio-fertilizers used in sustainable sugarcane crop production

7.3.1. Biofertilizers

Biofertilizers described as Preparations containing latent / living cells of various strains of microorganisms that allow crops to absorb or uptake of nutrients by their rhizosphere interactions when applied to seed or soil or seedling treatment. They accelerate other microbial processes in the soil that increase the amount of nutrients available in a form quickly assimilated by plants.

Biofertilizers are of different types

7.3.1.1. Nitrogen fixing biofertilizers

Table 14: Different N-fixing microorganisms used for biofertilizers

S.No.	Group	Microorganisms
N-fixing bacteria		
1.	Free living	*Azotabacter chroccocum, Klebsiella pneumonia, beijerinkia* sp. *Clostridium, Anabeana* sp.
2.	Symbiotic	*Rhizobium* sp., *Frankia* sp., *Anabeana* sp.
3.	Associative	*Gluconobacter diazotrophicus., Azospirillium lipoferum*

Azospirillum sp. and *Azotobacter* sp. *Azospirillum* sp. is more effective in clay soils with adequate nitrogen while *Azotobacter* sp. in sandy loam soils. Biofertilizer *Azotobacter* sp. and *Azospirillum* sp. saves about 20-25 percent of Nitrogen per hectare.

7.3.1.2. Phosphorus solubilizing

Micro-organisms like *Pseudomonas striata, Bacillus polymyxa, Aspergillus awamori, Bacillus magatherium,* etc. can solubilize the insoluble form of phosphate to soluble form. Application of phosphate solubilizing cultures increased the phosphate uptake and saved super phosphate equivalent to 30-35 kg P_2O_5/ha and also increases available phosphorus approximately 10 percent.

Table 15: Phophate solublizing biofertilizers

S.No.	Group	Name of microorganism
Phosphorus solubilising Biofertilizers		
1.	Bacteria	*Bacillus megterium, Bacillus subtilis, B. Circulans, Pseudomanas striata*
2.	Fungi	*Penicillium* sp. *Aspergillus awamori*

7.3.1.3 : Potassium mobilizers

Table 16: Potassium mobilizing biofertilizers

S.No.	Group	Name of microorganism
1.	Bacteria	*Fraturia aurantia*
2.	Arbuscular mycorrhiza	*Glomus* sp. *Gigaspora* sp. *Acaulospora* sp. *Scutellospora* sp. *Sclerocystis* sp.
3.	Ectomycorrhiza	*Laccaria* sp. *Amanita* sp. *Boletus* sp.
4.	Ericoid mycorrhiza	*Pezizella* sp.

7.3.1.4. Cellulolytic

In addition to these, many cellulolytic fungi can decompose recalcitrant crop residues, which builds up soil organic matter besides N-addition. With the intensive cultivation, deficiency of micronutrients viz., Zn, Mn, Fe and Cu is increasing at an alarming rate. Iron chlorosis can be corrected.

7.3.1.5. Biofertilizers for micronutrients

Bacillus sp. is a silicate and Zinc solublizing biofertilizers can be used in sugarcane crop.

7.3.2. Important biofertilizers used in sugarcane

7.3.2.1. *Azotobacter*

Azotobacter crococcum is an important species which are generally used in sugarcane crop as a biofertilizer. This species is multiplied in soil but not forming knots. *Azotabacter* convert the nitrogen into ammonia. This ammonical form is available to the plants for their nutrition. It also helps in improving germination, controlling the diseases and produces hormones for plant growth and development. *Azotobacter* saves 10-30 kg Nitrogen per hectare and increases sugarcane yield about 10-20 %.

7.3.2.2. *Azospirillium*

This biofertilizer is prepared from *Azospirillium brassilense* or *Azospirillium cipoferum* strains. It is found in nearby roots and saves about 30-35 % nitrogen. This is very helpful in increasing yield of sugarcane.

7.3.2.4. *Acetobacter*

Acetobacter diazotrophicus is a gram negative bacterium which is present in sugarcane crop (root, shoot and leaf). This helps in increasing yield of sugarcane and sugar about 5-20 tonnes per hectare and 5-20 percent, respectively. It also increases plant height, cane diameter. Acetobacter is an aerobic nitrogen fixing bacteria which increases the nitrogen at 30 percent sucrose concentration and *p*H (3).

7.3.3. Application of biofertilizers

About 4-5 kg of biofertilizer is required for one hectare land. Biofertilizer should be mixed with 100 kg compost or Farm yard manure (FYM) at the time of planting of sugarcane or at 30-45 days after sowing in the roots of sugarcane. Biofertilizers should be applied twice in a year. About 2-3 kg of *Gluconoacetobacter diazotrophicus* should be applied during hand hoeing and earthing up operation.

7.4. Rotation with legume crops

Sugarcane being a heavy feeder of nutrients requires huge quantity of fertilizers. So, to reduce the cost of fertilizers, diversification or rotation of sugarcane with pulse or legume crops can be a viable option because pulse or legume crop adds nitrogen to the soil and in addition to this, it also supplies phosphorus, micronutrients along with organic matter resulting into improved soil fertility and productivity. Legume crops like sunnhemp with spring planted sugarcane and Fenugreek, Berseem with autumn planted sugarcane or pulse crops such as French bean, Gram or Bengal gram, Pea, Lentil, Lathyrus, can be grown in autumn planted sugarcane and moong bean and cowpea in spring planted sugarcane. Residues of pulse/legume crops after harvest should be retain in the sugarcane field as mulch/ incorporated in the same field.

It will not only help in improving the soil physical, chemical and biological properties but also increases infiltration rate, reduces run-off and evaporation loss from the soil.

7.5. Use of industrial by-products of sugar mills

Press mud is an important by-product of sugar mills. Likewise, spent wash is also a by-product which is produced by distillaries of sugar mills. Both are important for sugarcane cultivation. Dry pressmud and Bio-compost are main source of nutrients which helps in improving soil physico-chemical and biological properties of soil. In sugarcane about 20 tonnes per hectare pressmud or bio-compost should be applied for improving quality of sugarcane juice. Use of Bio-methanated distillary effluents at the rate of 150m^3 /ha helps in increasing germination, crop establishment, tillering and development of sugarcane.

7.6. Integrated nutrient management

Sugarcane require huge amount of nutrient than other crops. For achieving higher productivity, farmers are using higher quantity of chemical fertilizers which leads to nuisance to the environment as well as soil. Organic carbon level is also diminishing day by day. So, there is a need to adopt improved techniques which are helpful in sustainable sugarcane production. Integration of organic (Manure, Vermicompost, Green manure, Bioferilizers) and inorganic sources (Chemical fertilizers) of nutrients can be a viable option for increasing production per unit area.

7.7. Crop logging in sugarcane

In Hawaiian Islands, H.F Clement conducted crop logging for analysis of plant nutrients in sugarcane. The crop log includes a set of chemical 23 (14/16) and physical measurements and is a graphical record of crop growth. These parameters indicate the general condition of the crop and suggest changes in management that are necessary to produce maximum yield. In the crop log method, critical approach to nutrient concentration is applied. The plant is cultivated in sugar cane for 35 days and analyzed for N, starch, sugar, moisture and weight of young sheath tissue. Nutrients like P and K are monitored at critical stages. Based on moisture and nutrient content, irrigation is scheduled and fertilizer application is done respectively and accordingly adjustment in management practices is followed as per need. By this record keeping, the productivity of crop is increased.

Chapter

8

IRRIGATION MANAGEMENT

8.1. Irrigation

It is the artificial application of water required for plant growth and development for ensuring crop productivity. Sugarcane is a water-intensive long duration crop of 11 to 18 months, with the ratoon crop up to three or more years. Too heavy rainfall is not beneficial for sugarcane growth and development results in low sugar content and drought during crop season produces a fibrous crop. Since vegetative growth is of economic part of sugarcane production, so there should not be any moisture stress in any growth stages. However, before harvesting of sugarcane crop, irrigation should be avoided as it deteriotes quality of juice. The total requirement of irrigation water in sugarcane crop is 1500-2000 mm annually.

Table 17: Average water requirements in the major sugarcane growing states of India

S.No.	Region/state	Water requirement (mm)	No. of irrigation (of 7.5 cm depth)
Sub-tropical India			
1.	Bihar	37.5	5
2.	Uttar Pradesh/ Uttarakhand	57.2	8
Tropical India			
3.	Andhra Pradesh	202.5	27
4.	Tamil Nadu	297.0	40
5.	Karnataka	256.0	34
6.	Maharashtra (Pre-Seasonal crop)	206.3	28
7.	Maharashtra (Adsali Crop)	243.8	33

Source: CWC, 2015

8.2. Water use and irrigation water requirement in sugarcane

Sugarcane is a water-intensive crop due to its long duration and accumulation of the largest biomass among all the crops. Maintenance of optimum soil moisture during all stages of crop growth is one of the essential pre-requisites for obtaining high cane yield under field conditions, water requirements are met by effective rainfall, the contribution from shallow water table and irrigation. It is estimated that about 80 percent of the irrigation requirements of sugarcane in India is met through groundwater sources. Frequent light irrigations, each of 40 to 50 mm, adjusted to suit the growing period of the crop and to the prevalent weather conditions are very useful. Water requirement is least in the ripening stage when the accumulation of sucrose starts and just before harvest irrigation is withheld for about a month. It has been estimated that the irrigation water requirement for sugarcane crop in India, ranges between 37.5 cm (Bihar) to 297 cm (Tamil Nadu) or about 5 - 40 irrigations of 7.5 cm each per hectare during its crop growth

In northern India, as the summer is hotter and drier than in southern India, the crop needs water more frequently. In central and western Uttar Pradesh about 8 irrigations are given which help the crop to tide over summer. In eastern Gangetic plains, cane subsists almost entirely on subsoil moisture and rainfall and receives around 5 irrigations at the most. The flat crop is irrigated by flooding; the crop on ridges/ after earthing-up receives irrigation through furrows. In initial stages, the crop can also be irrigated with sprinklers. Farmers in Maharashtra, Andhra Pradesh, Tamil Nadu and Karnataka have also adopted micro/ drip irrigation to economize on water application and achieve higher productivity and water and fertilizer use efficiency.

Table 18: Total consumptive water use and physical water productivity of sugarcane in the five production-wise groups for the dominant sugarcane producing districts of India

	Production-wise groups	Total consumptive water use (BCM)	Percent of total consumptive water use	Physical water productivity (kg/m^3)
I.	0-20 per cent	12.92	22.49	4.59
II.	20-40 per cent	12.64	22.02	. 4.95
III.	40-60 per cent	10.21	17.78	5.99
IV.	60-80 per cent	10.65	18.54	5.55
V.	80-100 per cent	11.01	19.17	5.25
Total/ Average		57.42	100.00	5.22

Source: CWC, 2014

Water should be given by following different criteria

8.3. Irrigation scheduling

Scheduling of irrigation means when and how much water to be applied for obtaining maximum production which is the pre-requisite for efficient use of irrigation water and production inputs like fertilizers, etc. The main objective of scheduling irrigation is achieving higher production per unit area with maximum water productivity and water use efficiency. Criteria for scheduling irrigation vary from place to place. In a situation, where rainfall is scanty or assured irrigation facilities and cheap water resources available, scientific irrigation scheduling should be based on soil-water-plant-atmosphere relationship. Irrigation scheduling decisions are to be made for mainly two conditions:

- Where water is expensive, irrigation should be scheduled to maximize crop production per unit of applied water.
- Where good land is scarce than water, irrigation should be scheduled to maximize crop production per unit of planted area.

There are various approaches which are used for scheduling irrigation in sugarcane are as follows

8.3.1. Critical growth stages

Sugarcane crop has certain critical stages at which if there is shortage of moisture in soil, yield and productivity will reduce abruptly. Water requirement in sugarcane crop is different at different growth stages of crop. When water is limited, it is necessary to schedule irrigation according to the critical growth stage in sugarcane. Critical stage for irrigation in sugarcane is germination (0-45 DAP), formative (tillering) at 60- 130 DAP and grand growth stage at 130-250 DAP because water requirement is maximum in these stages. Approximate water requirement of 12 month sugarcane crop at each growth phase is indicated in Table 19.

Table 19: Water requirement of sugarcane crop at different growth stages

S.No.	Growth stages	Water requirement (mm)
1.	Germination	300
2.	Tillering stage	550
3.	Grand growth stage	1000

8.3.2. IW/CPE approach

This is climatological approach based on Evapo-transpiration (ET) and helpful for scheduling irrigation. It depends upon ratio of IW/CPE (Irrigation water/ cumulative pan evaporation). Broader ratio indicates no need of irrigation and vice versa. The amount of irrigation given is either equal to ET or fraction of ET. In IW/CPE approach, a known amount of irrigation water (IW) is applied when cumulative pan evaporation (CPE) reaches a predetermined level. The amount of water given at each irrigation ranges from 4 to 6 cm, most common is 5 cm. In general, scheduling irrigation at IW/CPE ratio of 0.75 is ideal in sugarcane, if there is no scarcity of irrigation water.

8.3.3. Soil moisture depletion approach

The moisture between field capacity and permanent wilting point forms the basis for soil moisture depletion approach. Irrigation in sugarcane at 50 percent depletion in available soil moisture (DASM) in the root zone at 15-30 cm soil depth during active growth stages and 75 percent depletion during maturity leads to maximum crop yield. However, scheduling irrigation in sugarcane at 50 percent DASM is ideal for higher production.

The moisture deficit may be calculated by using following formula.

$$d = \frac{FC - A}{100} \times BD \times D$$

Where, d is moisture deficit in a soil layer

FC is Field capacity (%)

A is actual moisture content of soil

BD is Bulk density of the soil (g/c.c)

D is depth of soil layer in cm

8.4. Method of irrigation

8.4.1. Flooding

Flooding irrigation is best used for row crops such as sugarcane. This method is followed in smooth and leveled sugarcane fields and where water is in abundant quantity and inexpensive. There is no control over the flow rate or distribution and wastage of water leads to the low water use efficiency.

8.4.2. Furrow method

This method is suitable for sugarcane crop. In this method, water is applied in the furrows which were developed between crop rows in the planting and cultivation processes. There is usually one furrow between each row of crop. Furrows are laid down along the slope when slope of the land is about 3 percent. The furrows are laid on graded contour when the slope exceeds 3 percent and up to 15 percent. The length of furrows varies with soil type, slopes of the land and the quality of water to be applied and may vary from 10-100 meters.

Advantages of furrow irrigation

1. Furrow method is suitable for those crops which are sensitive to saturated soil condition in the root zone.
2. Water in the furrows contracts only one fifth to one half of the land surface, thereby reducing puddling, crushing of the soil and evaporation losses.
3. Uniform application of water

Disadvantages of furrow irrigation

1. This method is not suited for canal areas
2. Higher cost involved in the preparation of furrow

Following are the different types of furrow irrigation

1. Straight furrow method of irrigation
2. Contour furrow method of irrigation
3. Corrugated furrow irrigation

8.4.3. Drip irrigation

It is also known as trickle irrigation. The modern technique drip irrigation was developed by Simcha Blass in Israel. Now a days, drip irrigation is popular method of irrigation in sugarcane crop because in this method, water is applied in precise quantity but slow application as discrete continuous drops, tiny stream through mechanical devices called emitters/ applicators located at selected points along with water delivery lines and saves water upto 40-70 % and increases water use efficiency and water productivity. However, the initial cost of drip irrigation is high but in sugarcane growing countries it is widely used. In India (Maharashtra) irrigation through drip method is very popular. Government is providing subsidy for drip irrigation to sugarcane growers. Drip irrigation is of two types.

(a) Surface drip irrigation

The application of water to the soil surface as drops/ miniature through emitters placed at pre-determined distance along with laterals.

(b) Sub-surface drip irrigation

The application of water to the soil surface as drops/ miniature through emitters molded on the inner wall of drip line.

Advantages of drip irrigation

1. Increased beneficial use of available water
2. Enhanced plant growth and yield
3. Reduced salinity hazard to the plant
4. Improved fertilizer use efficiency due to fertigation in drip system
5. Limited weed growth
6. Reduced operational labour
7. Improved cultural practices

Disadvantages of drip irrigation

1. Initial higher cost and high maintenance cost
2. Salt accumulation near the plants
3. Economical and technical limitations

8.4.4. Sprinkler irrigation/over head method of irrigation

In some sugarcane areas, water is applied by means of sprinkler irrigation. In this method, water is sprayed in the form of rainfall. Sprinkler irrigation is applicable in the areas where elevation is higher and land is undulated and sandy soils requiring excess application with surface methods. Application efficiency of overhead irrigation is about 60-65 percent. The advantages of this system are elimination of channels, bunds and their maintenance and conveyance losses of water and controlled water application leading to higher efficiency. Only limitation is its higher initial cost, high maintenance requirements, and higher energy cost and not effective in the areas where wind speed is high (> 18 km per hour).

8.5. Drainage

Drainage is also very important in sugarcane crop. Excess water should be removed from sugarcane field. Sugarcane juice quality will reduce drastically due to water logged conditions. However, now a days there are so many varieties which can withstand waterlogged conditions up to more than 7 days.

Chapter

9

WEED MANAGEMENT

9.1. Weed

Weed is a plant growing where it was not desired given by Jethro Tull (1731). Weeds create problems in crops and interfere with the growth and development of crops. Weeds are more problematic in the case of sugarcane because of its long duration, wider spacing and slow initial growth habits.

It is cultivated in a different time, so weed flora is not common for all planting season, it varies from place to place and season to season and weed incidence depends upon soil type, climatic conditions of the region. Some common weeds found in autumn and spring-planted sugarcane are as follows.

Important weeds in autumn planted sugarcane

Narrow-leaved weeds: *Phalaris minor* (Gulli Danda), *Avena fatua* (Jangal Jai) etc.

Broad-leaved weeds: *Convolvulus arvensis* (Bind weed) *Ipomeas spp*. (Morning glory) *Chenopodium album* (Fat hen/ Bathua), *Cirsium arvense* (Canada thistle), *Euphorbia hirta*, (Badi dudhi), *Melilotus alba* (Senji), *Medicago dentriculata* (Medick/burclover), *Oxalis corniculata/latifolia* (Sorrel), *Anagallis arvensis* (Krishnil), *Amaranthus viridis* (Jangli chaulai), *Solanum nigrum*, (Black night shade) *Physalis minima* (Ground cherry), *Vicia sativa* (Akari), *Canabis sativa* (Hemp) etc.

Sedges: *Cyperus rotundus* (Motha), *Fimbrystylis* spp. (Hoorah grass)., *Fumaria parviflora* (Haussk),

Important weeds in spring planted sugarcane

Narrow leaved weeds: *Cynodon dactylon* (Doob grass), *Digitaria sanguinalis* (Crab grass), *Elusine indica* (Goose grass), *Sorghum helepense* (Johnson grass), *Dactyloctenium aegyptium* (Crow foot grass)

Broad leaved weeds: *Ageratum coinozoides* (Goat weed), *Rumex* Spp. (Curlydock), *Ipomea* spp. (Morning glory), *Eclipta alba*, *Digera arvensis*

Sedges: *Cyperus rotundus* (Purple nut sedge), *C. iria* (Rice flats edge), *C. difformis* (Small flower umbrella plant), *C. esculentus (Yellow nuts edge).*

9.2. Yield and quality reduction in sugarcane

Weeds compete with the crops for space, light, nutrient, water and other resources resulting into reduced growth and development. Researches shows that yield loss in sugarcane is approximately 17.5 to 35.4 t/ha. Crop quality is also reduced due to weed competition in the sugarcane field. Twining weeds such as *Ipomoea sp*. that twines clump together affect cane growth and cause yield losses of about 25 per cent. Twining of weeds increases cultivation costs and causes serious consequences. The reported cane yield losses in India range from 12% to 72%. If weeds in the initial stages are not properly controlled, the yield loss could be as high as 17.5 t / ha. Poor sugar cane growth caused by weeds also infestation causes deterioration in quality. Cogan/bermuda grass (*Cynodon dactylon*). Grass (*Imperata cylindrica*) is graminacious weeds are known to act as hosts of Ratoon Stunting Disease (RSD) of cane sugar. Weeds remove nutrients such as 4 times Nitrogen N and P, 2.5 times of K compared with first sugar cane during seven weeks period. The degree of competition is dependent on (a) weed flora, (b) time of weed emergence (c) relative density of crop plants and weeds, (d) duration of competition, etc. Competition is set in motion when the supply or availability of resources is limited for the unrestricted growth of both crop and weeds.

9.3. Critical period of crop-weed competition (CWC)

CWC is very important in weed management. The critical period of weed competition is the shortest time span during the crop growth when weeding results in highest economic returns. The critical period of weed competition is approximately 1/3rd of the duration of the crop. CWC play crucial role for each and every crop. If weeding is not done at this period crop yield drastically reduced.

Cyperus rotundus (Purple nut sedge)

Rumex acetacella (Jangle palak)

Circium arvense (Canada thistle)

Chenopodium album (Bathua)

Blind hoeing in sugarcane crop infested with Doob grass (*Cynodon dactylon*)

Hand weeding

Figure 6: Important weed flora in cane crop, sugarcane based intercropping system and cultural operations in sugarcane fields for weed control

9.4. Management options

Effective weed management is most important in the early stages of crop development. There are so many options of weed management in sugarcane. These are as follows

9.4.1. Physical methods

The physical control of weeds involves the use of manual or mechanical energy for destroying weeds. It is oldest, but most common method of weed control.

9.4.1.1. Hand weeding

It is a very effective method of weed control; it removes every blade of unwanted grass growing among the crop plants. Hand weeding is generally done by using sickles, axes and spades. Hand weeding is a very slow and tiresome method of weed control.

9.4.1.2. Hand hoeing and tillage

The primary function of any tillage operation is to control weeds although, besides it, it may also achieve some other objectives like the incorporation of manures and fertilizers into the soil; breaking of hard pans, forming ridges, etc.

Pre-plant tillage is usually conducted in two phases, viz., the primary tillage and secondary tillage. The primary tillage is conducted effectively with either a soil inverting plough or disk plough.

The secondary tillage, on the contrary, is performed with lighter implements like harrows, cultivators, weed-mulchers and corrugated rollers. The primary tillage can help in the control of weeds by:

1. Burying the existing weeds, this in turn, eases the post-plant tillage and the application of pre-plant herbicides.
2. Bringing the weed seeds to soil surface for germination and their subsequent destruction by suitable, secondary tillage operations.
3. Incorporation pre-plant herbicides
4. In case of perennial weeds, delaying the need of post-plant tillage.

The role of secondary tillage in weed control is mainly to dislodge the germinating weed seedlings.

9.4.1.3. Mowing

Mowing is cutting of a uniform growth of weeds from entire area at the ground-level. In certain cases, to prevent multiplication of weed seeds. Also, repeated mowings can weaken the underground parts of perennial weeds. Mowing is usually practiced in non-crop areas, lawns and gardens, but it can also prove useful in removing weeds from rows of certain established crop such as sugarcane, potato, etc. Common mowing tools and implements are sickle, sword, reciprocating type rotary-bar mower. Mowing is effective against only erect, herbaceous weeds.

9.4.2. Cultural methods

This is old practice of weed control. Good crop husbandry is an inevitable component of practical weed management plan in any agricultural system. Some important crop husbandry practices that can lead to the suppression of weeds, and which can be easily incorporated in practical weed control programmes in farming systems are discussed in this chapter. The approach is sometimes also referred to as ecological management of Weeds.

9.4.2.1. Proper crop stand and early seedling vigour

Uniform germination, good crop stand and vigorous crop seedlings of sugarcane leave less space for growth and development of weeds. On the contrary, low plant population and weak crop seedlings permits weed infestation in the sugarcane field. So, keeping a proper crop stand is necessary for controlling population of weeds. Unnecessarily deep planting of sugarcane should be avoided as it results in slow germination.

9.4.2.2. Proper planting method, time and crop density

Proper planting method and time is very important for reducing weed growth. By adjusting the sowing period in such a way either to sow crops before germination of weeds or after emergence of weeds which can be uprooted by repeated ploughing.

The sugarcane crop due to its wide spacing and long duration facilitates more weed infestation. So, proper time of planting is necessary for removal of weed flora.

9.4.2.3. Crop rotation

Crop rotation is the growing of crops in rotation in a year. Example– Rotation of sugarcane crop with pulse and oilseed crops. Rotation of crops having dissimilar life cycle breaks the continuity of weeds due to changes in ecological situations. Crop rotation minimizes the weed infestation. It is best non-monetary input of controlling weed.

9.4.2.4. Summer fallowing

Fallowing is an old practice of controlling weeds. The purpose of fallowing is to destroy weeds by damaging them through cultivation and allow it to dry. Summer fallowing is more beneficial for reducing weed problems because of high temperature and soil moisture. It is effective method of controlling perennial weeds such as *Cynodon dactylon*.

9.4.2.5. Intercropping

Sugarcane intercropping with pulses, oilseeds, millets, fodders, spices, vegetables, medicinal and aromatic crops reduces the weed population to a greater extent. Intercropping with different crop groups is a potential weed management practice especially in crops with wider spacing such as sugarcane, potato, etc. Crops which have quick growth is very helpful in the sugarcane intercropping for smothering weeds, reducing crop-weed competition. Commonly adopted crop in sugarcane intercropping are cowpea, coriander, berseem, chick pea, tulsi, garlic, onion, french bean, cabbage, etc.

9.4.2.6. Flooding and Drainage

Flooding in sugarcane is very effective tool in controlling sedges, broad leaved weeds in sugarcane crop. At the same time, drainage is equally important for controlling creepers like *Ipomea* sp.

9.4.2.7. Green manuring

Closely sown green manure crop and its quick growth suppresses the weed population in a significant manner and when green manure crop is turned into the soil, weeds also buried in the soil and helps in increasing organic matter. Growing of green manure crops during summer followed by fallowing in the month of July-November is the best possible way of controlling weed infestation in the autumn planted (Nov-Dec) sugarcane fields.

9.4.2.8. Nutrient management

Better nutrient management with 4 R principles (Right time, Right place, Right method, and Right quantity) eliminates the weeds in sugarcane crop to a greater extent. Placement of fertilizers closer to the root zone of crop plants helps in restricted use of fertilizers by weeds. Band placement of fertilizer is better than the broadcasting method in sugarcane cultivation.

9.4.3. Preventive measures

Weeds have to be controlled efficiently in order to reduce negative implication of weeds on sugarcane crop growth and productivity. Weed prevention comprises all measures which delay the entry and establishment of weeds in a new area where it was not introduced earlier. Principles of weed prevention are as follows.

1. **Use of well decomposed manure**

   Partially decomposed manure or compost contains weed seeds which have longer viability can be disseminated through application in the sugarcane field. Therefore, it is advisable that always use fully decomposed organic manure / FYM/ Compost.

2. **Use of cleaned machinery and devices in sugarcane fields**

   Before primary, secondary tillage and post-harvest operations, machinery and implements should be cleaned properly, so that contamination of weeds in the sugarcane field can be minimized.

3. **Prevent movement of weeds with other farm resources**

   Do not permit livestock to move from the weed infested areas into cropped areas because they can disseminate weed seeds and fruits which is stuck to them or ingested by them earlier.

9.4.4. Chemical methods

Chemical method is the easy way of weed control in sugarcane and extensively used in India as well as world because of labor scarcity, inefficient conventional methods, etc. However, this method is very effective method of weed control. It is important to note that the growers understand well about the impact of adverse weather, that is, heavy weather conditions rain, on the timing of herbicide application. A weed-free environment is absolutely essential. Herbicides are applied at different times in sugarcane depending on their mode of action and accordingly grouped as follows:

(1) **Pre-planting herbicides:** The herbicides applied before planting of cane crop in the field. Pre-planting herbicides are non-selective. The herbicides need to be applied into the soil and incorporated properly as they are volatile in nature. This type of herbicides affects the weeds seedling and act as a contact killer.

(2) **Pre-emergence herbicides:** The herbicides that are applied after the planting of sugarcane but before emergence of weeds. Most pre-emergence herbicides are absorbed in the surface layer of the soil where they can kill seedlings as they emerge through the soil layer. Pre-emergence herbicides provide an ideal weed control in sugarcane crop.

(3) **Post-emergence herbicides:** The herbicides that are applied only after the planting of sugarcane and emergence of weeds. The herbicides applied on

sugarcane crop at the seedling and tillering stage at about 30-60 days after planting.

Table 20: Herbicides used in sugarcane

Herbicide application	Herbicide (trade name)	Dose (kg or g or litres a.i./ha)	Time of application	Weed flora
Pre-plant incorporation (PPI)	Fluchloralin (Basalin) / Trifluralin (Treflan)	1.0-1.5 kg a.i. /ha	Before planting	All weed flora
Pre-emergence application (PE)	Pendimethalin (Stomp Xtra)	2.0 kg a.i/ha	4-5	Narrow leaved weeds
	Atrazine (Atrataf)	1.75kg a.i. /ha	3-4	Narrow leaved and broad leaved weeds (BLW)
	Oxyfluorfen (Goal)	0.75 litres a.i./ha	15 or 30	Grasses/BLW
	Diuron (Karmex)	1.6-3.2 kg a.i./ha	10 or 15	Grasses/Sedges
	Metsulfuron Methyl with non-ionic surfactant	6 g a.i/ha	15 or 30	All weed flora
	Hexazinone (Barrage) + Diuron (Karmex)	1200 g (264+936)	15 or 30	All weed flora
	Ethoxysulfuron	80 g a.i/ha		
Post-emergence application (PoE)	Gramaxone + 2,4-D Sodium salt	2.5 litres a.i./ha + 2.5 litres a.i./ha	21	Broad leaved weeds (BLW)
	Metribuzin (Sencor) + Dicamba (Cadence)	1.25 kg a.i./ha + 350 g a.i./ha	60	Creeper weed like *Ipomea* sp.

9.4.5. Biological methods

Many scientists used living organisms such as insects, microorganisms, herbivorous fish, pathogens and competitive plants for controlling of weeds called as Biological control. These bio-agents are host specific which should not attack other plant species. In sugarcane weeds like *Cyperus rotundus*, *Parthenium hyterophorus*, *Sorghum helepense* can be controlled by biological means. *Bactra verutana* is a shoot boring moth can be effective in controlling *Cyperus rotundus*. Carrot grass/ Congress grass can be controlled by using Mexican blue beetle, *Zygogrammma bicolorata* at the rate of 50-100 beetles per plant of carrot grass. *Sorghum helepense* (Johnson grass) weed can be effectively controlled by using mycoherbicide Bipolaris (A suspension of fungal spores of *Bipolaris sorghicola*), etc.

9.4.6. Integrated weed management in sugarcane

Integrated Weed Management (IWM) embodies a grower's knowledge of the physical characteristics of Farm, identifies the species and densities of weed and products and equipment ready for use. Use of all appropriate means of weed control together to maintain the weed populations under the degree of economic damage is known as integrated weed management (IWM). Being a long duration and wide spaced crop, there is considerable scope to use tools cultural / mechanical and chemical combined to reduce reliance on any of these approaches.

Why integrated weed management is necessary?

1. One herbicide may or may not be effective to control the weeds
2. One method of weed control mayor may not be effective in controlling broad spectrum of weeds
3. Continuous use of herbicides can develop resistance in weeds against particular chemicals
4. Due to environmental protection

IWM

Cultural methods	Physical methods	Chemical methods	Biological methods
1. Deep tillage 2. Crop rotation 3. Summer fallowing 4. Green manuring 5. Proper planting time, method & plant population 6. Intercropping 7. Flooding & drainage 8. Nutrient management 9. Soil solarization 10. Crop competition	1. Hand hoeing & tillage 2. Hand weeding 3. Mowing 4. Digging, burning	1. Herbicides use • Pre-plant incorporation • Pre-emergance application • Post-emergance application	1. Biological agents 2. Mycoherbicides

❖❖❖

Chapter

10

INSECT PEST MANAGEMENT

10.1. Borers pest

10.1.1. Shoot Borer, *Chilo infuscatellus* Snellen (O-Lepidoptera: F- Crambidae)

It has a wide range of occurrence from Afghanistan through central Asia, India to Korea, Taiwan, and Indonesia as well as Philippines. The shoot borer is widely distributed in all sugarcane growing areas in India, infesting the crop during its early stages of growth in spring or autumn planted crop. It occurrence of pests are during the period March to June. It infest heavily during drought or scanty rainfall. High temperatures and low to moderate humidity favored borer activity.

Damage Symptoms

- Central whorl of leaves dries up and forming dead hearts.
- Dead hearts of infested plant can be pulled out easily.
- Rotten portion of straw emits offensive odor.
- A number of feeding punctures near the base of shoot.
- The infection starts just after germination and continue up to June
- Borer attack during the germination stage kills the mother shoots.
- Resulting in the drying up of the entire clump, creating gaps in the field.

Figure 7. Damage shoot

Figure 8. Punctures near the base of shoot

Life Cycle

Moths are light straw to brownish colored and nocturnal in nature. It laid eggs in clusters on the undersurface of middle portion of the first three leaves. It take time for hatching in 4-6 days then the larva wanders for a few hours and reaches the base of the stem by crawling or hanging by silken thread. Larvae moult 4-5 times passing through 5-6 instars. Body is cylindrical and exhibits a dirty white colour with 5 violet dorsal stripes from second thoracic to eighth abdominal segment. Larval period lasts for 16-30 days. Larva pupates within the plant. The pupae are elongate, slender and yellowish to dark brown in colour. Pupal period lasts for 6-12 day.

Management

- Early planting should be followed to avoid synchrony between the susceptible crop stage and peak borer activity period.
- Two to three light earthing-up and trash mulching operations should be given during the early stages of the crop growth.
- Removal of dead heart at fortnightly interval.
- Intercropping of sugarcane with coriander, onion and garlic, potato and green gram, depending on planting season, reduces shoot borer incidence.
- Installed pheromone traps one month after planting @ 25/ha with monthly change one lure help to reduce moth population.
- Egg parasitoid Thichogramma chilonis sequential releases @ 50,000 /ha at 10 days interval beginning 30 days after planting during April-June would be useful.
- Severe condition may apply cartap hydrochloride 4G @25kg/ha followed by light irrigation.

10.1.2. Root Borer, *Emmalocera depressella* (O-Lepidoptera:F- Pyralidae)

This is the only species of borer complex of sugarcane infesting the underground part of canes and so called as 'root borer'. It infests in the north Indian sugarcane belt, Northern areas of Gujarat, Maharashtra, Karnataka and Andhra Pradesh. It also occurs in Pakistan. Malaysia and Bangladesh. It is a serious pest in the subtropical region

DAMAGE SYMPTOMS

Damage Symptom of root borer

- It attacks the underground shoot portion of cane in the early stages, i.e. during May-June.
- Infested cane of underground part observed a single fine hole at the base of shoot and one or two open leaves may also dry up along with the dead heart.

Figure 9. Damage Symptom of root borer

- Dead hearts are formed which cannot be easily pulled out and do not emit offensive smell. Often one or two leaves adjacent to the central whorl also dry-up.
- In grown up plant, the damage symptom is observed in the form of yellowing of leaves.
- Yield loss and sucrose reduction have been observed due to attack of this pest.

Life Cycle

Forewings of adult moths have straw colored with whitish hind wings. Female lays egg singly on the lower surfaces of leaves and difficult to detect. After an incubation period of 5-8 days the first instar larva crawls to the base of the stem and enters it by making a single hole at or just below the ground level. Larval period lasts from 20 to 45 days in summer and longer during rainy season and winter. Larvae hibernate during October-February. The full grown white colored larva cuts an opening at the exterior end prior to pupation and constructs a silky pupal tube up to the surface of soil for easy emergence of the adult. Pupal period lasts for 7-11 days during summer and rainy seasons and 8-14 days during winter. Life cycle is completed in 30-73 days depending on the season and region. The borer is active at high temperatures and moderate humidity levels. Generally ratoon is more prone to its attack. Unirrigated fields and sandy or sandy loam soils harbor high borer incidence.

Management

- Collection and destruction of affected shoots along with borer larvae.
- Deep harvesting of canes that art to be ratooned should be adopted to expose larvae or pupae.
- Avoidance of ratoons if field heavily infested to minimizes proliferation of the pest
- In cases of severe infestation, fields should be ploughed up and affected stubbles with larvae should be collected and destroyed.
- Collection and destruction of moths using light traps.
- Release of *Trichogramma chilonis*@50,000/ha at 10 days intervals from March to June.

10.1.3. Top Borer, *Scirpophaga excerptalis* Walker (O-Lepideptera: F-Pyralidae)

It is unique among the borers pest and infests in sugarcane during all the stages of crop growth. It occurs as a regular and major pest in the subtropical region and It is active from March to November. Maximum damage is occurred during April to July as it feeds on sugarcane.

Damage Symptoms

- Due to attack of 1st and 2nd brood of top borer during I earlier stages of sugarcane plants have showed reddish streaks together with small holes in the plants.

- The young larvae bore into the leaf through midrib and consequently the main shoot or cane stalk.
- Formed deadhead but cannot be pulled out easily.
- In grown up canes due to the infestation dead heart formation induces sprouting of the lateral buds giving a 'bunchy top' appearance.

Figure 10. (a) adult, (b) egg masses, (c) damage symptom and (d) Bunchy top

Life Cycle

Moth is silvery white in color with or without one black spot on each of the forewings. The female has a toll of crimson red, orange or buff colored anal hair. Eggs are laid in clusters of two to four overlapping rows covered with buff colored hair mostly on the underside of the basal or middle portion of second and third lamina. . An egg mass may have 10-80 eggs and a female lays 46-216 eggs during its life span of 1-4 days. Incubation period varied from 6-13 days and newly hatched larvae crawl actively on leaves. This 1st instars larvae mining for 1-2 days and they come out of the midrib only through the upper epidermis and enter into the interior of the spindle and continue to tunnel through the soft midribs. After spending 15 - 45 days in four instars, the full grown larva cuts an exit hole for the moth that would emerge after a papal period of 6-21 days.

Management

- Planting on time may be altered wherever the borer is severe during these months.
- Intercropping with spices reduces top borer incidence.
- Setting up of sex pheromone traps fie 50 numbers/ha reduces moth population.
- Hand collection and destruction of egg masses is effective to reduce population of insect.
- Isotima javensis is an efficient natural parasitoid should be release in problem areas.
- The egg parasitoid *Trichogroma japonicum*@50000/ha released 4 to 6 times at interval of 10 days during second and third broods was found to reduce damage of this borer.
- In severe cases of infestation, Cartap hydrochloride 4G @25kg /ha or Fipronil 0.3 G@ 33 kg/ha may he applied in the root zone, covered with soil followed by irrigation against third brood in the third week of june.

10.1.4. Internode Borer, *Chilo sacchariphagus* Indicus (O-Lepidoptera: F-Crambidae)

The internode borer is a important pest of sugarcane in peninsular region of India. It was reported from south India. It is distributed in the slates of Tamil Nadu, Kerala, Andhra Pradesh, Karnataka, Maharashtra, Uttar Pradesh and Haryana.

Damage Symptoms

- The activity of this borer start just after internodes formation and the damage continues till harvest.
- The newly emerge larva feeds on the ligular region and inner side of leaf sheath by scraping the epidermis for seven to eight days.
- The larva tunnels the inner tissues of the intemode and pushes the frass to the exterior.
- The frass of damage cane can be seen as a wet or dry mass hanging from the bore hole which can serve as an indicator of borer attack as well as the duration of attack.
- Borer attack reduces weight of millable canes and yields of crop.

Life Cycle

Moths are straw colored with a light dark spot on each of the forewings. Males are smaller and darker in color than females. Female moths lay eggs in clusters of 2-3 parallel rows on both surfaces of the green leaves, dorsally on the midrib and ventrally parallel to the midrib. A single moth lays up to 414 eggs. The incubation period is five to six days. The larva moults six to seven times with a duration of 37-54 days. The grown up larva has violet spots in four longitudinal stripes dorsally.

The larva in the pre-pupal stage comes out of the tunnel and selects a semidry lead-sheath, spins a silken cocoon and pupates inside. Freshly formed pupa is light brown in color becoming dark brown after six to seven days. The pupa has a row of rough projections on fifth to seventh abdominal segments. Adults emerge in 7-10 days. Heavy buildup of the borer occurs during July to December after the commencement of internode formation. The factors like lodging of plant, excess application of nitrogen fertilizer, water logged condition and presence of large number of water shoots favor pest buildup.

Management

- Removal of water shoots at 8-9 month age of the crop helps to prevent perpetuation of this borer.
- Periodical detrashing at fifth and seventh months is effective for reduce incidence.
- Installed pheromone traps @ 25/hectare at age of fourth or fifth month of the crop with 4-5 lure changes at 30 day intervals reduces moth population.
- Release of *Trichogramma chlionis* @ 50000 Egg parasitoid/ha at 10 days interval recommended for internode borer control for a long time.
- Conservation of larval parasitoid Cotesia flavipes is effective for reduce incidence.

10.1.5. Stalk borer, *Chilo auricilius* (O- Lepidoptera- Crambidae)

It was first time recorded from Bengal in year 1905. Fletcher and Ghosh reported it as a pest of sugarcane at Pusa, Bihar. In some parts of Uttar Pradesh, it occurred in epidemic form in 1940. It is one of the most destructive pests of sugarcane crop in subtropical India and a major pest in Uttar Pradesh, Haryana, Punjab and Bihar.

Damage Symptoms

- The larvae feed on plant by scraping the inner surface of these leaf sheaths for a week and they complete two instars.
- Later they mine in one leaf sheath due to which the leaf sheath decays and subsequently the entire leaf dries.
- Due to continuous feeding of larvae leaves show distinct longitudinal orange yellow streaks extending from tip to base on both sides of the midrib towards leaf margins.

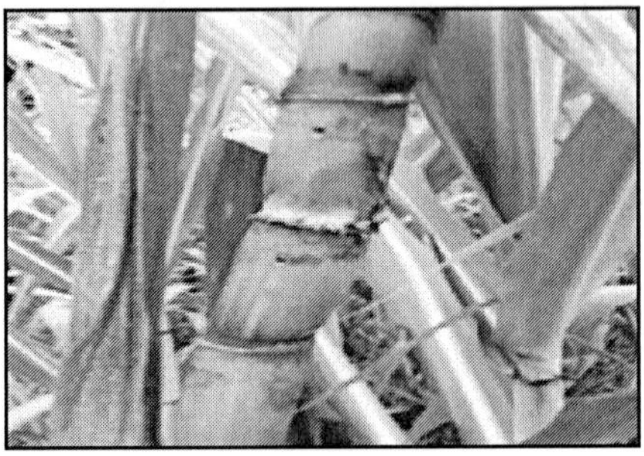

Figure 11. Infested internodes of Stalk borer

- The grownup cane of infested plant does not visible unless the leaf sheaths are removed when the bore holes may be sum clearly on the internodes.
- By splitting of damaged internodes show symptoms of reddening which sometimes emit a rancid odor.

Life Cycle

Moths have straw colored forewings with a number of golden spots on the periphery and light straw colored hind wings with silvery fringe. The eggs are laid in masses on the underside of the leaves arranged in two to five parallel rows. The incubation period ranges from 4-10 days. The newly hatched larva is creamy white in color which normally passes through five instars but in summer and winter there can be as many as eight instars. The mature larvae have five violet stripes along the length of the body. The larval period varied from 16-51 days and pupation takes place inside the tunnel which is formed by the larva. The pupal period lasts from 6-10 days during March to October and 20-24 days in January. The borer completes five generations in Haryana, five to six in Uttar Pradesh and seven in Bihar.

Management

- Detrashing of cane during September-October will expose the infested internodes to the activity of natural enemies.
- Removal of water and late shoots at fortnightly interval.
- Release of *Trichogramma chilonis* egg parasitoids@50000/ha at 10 days interval from July is useful in reducing the infestation level.
- Conservation of braconid parasitoid *Cotesia flavipes* help to reduced stalk borer infestation.
- Detrashing followed by application of systemic insecticides, Diamethoate 30EC @ 2lit. /ha twice during August and September with one month interval would be useful.

10.1.6. Gurdaspur Borer, *Acigona steniellus* (O-Lepidoptera: F-Crambidae)

It is a serious pest of sugarcane in some pockets of north India. Dutta observed, it as a major pest in Gurdaspur district of Punjab and named it as "Gurdaspur borer" in year 1925. Subsequently, Gupta and Garg recorded severe damage by this borer in the Dehradun district of Uttar Pradesh ad referred to it as "Dehradun" borer. It is also reported from Pakistan and Vietnam.

Damage Symptoms

- The pest exhibits both gregarious and solitary phases.
- In early stage, the larvae feed gregariously in the top portion of the canes by making spiral galleries thus killing the plant.
- Externally the tunnel appears as a dark spiral made up of a series of punctures.
- As many as 35 larvae may real gregariously in one cane.
- After feeding two thirds of internode in a spiral manner, the larvae bore deeper and feed upward by making a straight tunnel.
- Third instar larvae start dispersing to other canes singly or in groups of two to three and in the later stages one caterpillar is found infesting the internodes in each plant.

- Leaves wither due to larval feeding just below the rind but as the infestation advances and the central tunnel is formed, the entire whorl of leaves turns yellow and dries up.
- As a result of the attack, canes fail to grow, damaged internodes become weak and infested stalks break off easily.
- Nodal buds in the remaining portions of canes are stimulated and side shoots are formed.

Life Cycle

Adult females lay about 90-300 scale like eggs in clusters on the leaves along the midrib in 3-4 rows. They incubate within 4-9 days. Young larvae enter top portion of the cane through a single hole just above the node. They feed in groups by making spiral galleries which run upwards. After 7-10 days they attack the adjacent canes. The larval period varied from 21-27 days and they pupate inside the cane. The life cycle is completed within 35-40 days. Sugarcane tops are dry due to the attack of this borer during July to September. The pest flourishes well under moderate temperature and high humidity conditions. Early monsoon, heavy rains and water logging also favor its multiplication.

Management

- Affected canes with up to 5 to 6 top internodes harboring borer larvae in the gregarious phase should be cut and removed regularly.
- Larvae in stubbles can easily identified by the presence of round tunnel holes at the cut ends, should be uprooted and burnt to kill the larvae.
- Collect egg masses and destroyed them.
- Ratooning of heavily infested crop should be avoided to prevent perpetuation of the pest.
- Spraying Monocrotophas 36SL @2lit/ha for control of this pest.

10.1.7. Plassey Borer, *Chilo tumidicostalis* (O-Lepidoptera: F-Crambidae)

It was referred to as 'Plassey borer' by Gupta bemuse of its occurrence in an epidemic form at Plassey (West Bengal) during 1956. it is of economic importance only in Assam and Bihar. It also occurs in Nepal, Bangladesh and Burma.

Damage Symptoms

- The incidence of the borer can be easily recognized by the presence of dried crown of infested sugarcane.
- Borer attack occurs in two distinct forms.
- Primary infestation is caused by as many as 156 newly hatched larvae aggregating in the top three to five in internodes.

Figure 12. Damage symptom of plassey borer

- Due to such gregarious feeding, fresh red colored frass is seen projecting out of the bore holes and the top leaves of the canes ultimately dry up.
- Affected internodes break off easily and nodes adjacent to the infested internode develop sett roots which completely envelop the stalk; sprouting of nodal buds also occurs.
- In the case of secondary infestation, the grownup larvae disperse to either the lower healthy portion of the canes showing primary attack or the neighboring canes. Larvae feed by tunneling the internodes. One larva may bore into one to five internodes in a cane but cane tops do not dry up in this form of damage.

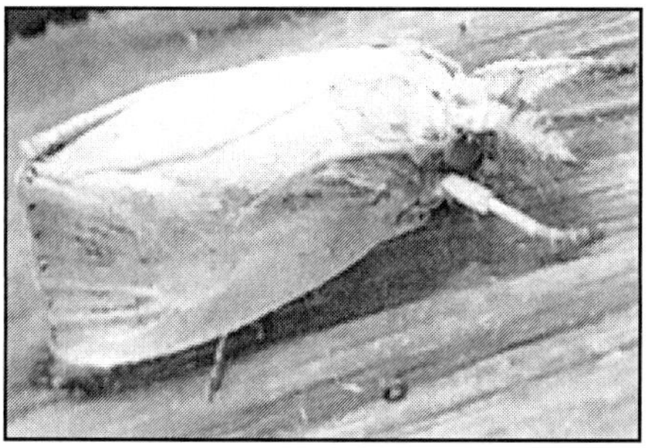

Adult

Larvae

Pupa

Infested internode

Figure 13. Different stages & insect with infested internode

Life Cycle

Moths are cinnamon brown with terminal series of black spots broken externally by small silvery white points. Hind wings whitish except for a few light brown scales in the coastal area of male moths. Egg masses are deposited on the underside of top one to three leaves in two to four tiers. Occasionally, eggs are also laid on leaf sheath and stalk. Larvae from one egg mass hatch synchronously in seven days and penetrate into one of the tender internodes. After 10 days of primary infestation, the larvae disperse to adjoining canes each one boring into a separate internode. There is a well marked polymorphism and larvae are of four different types. However, a typical larva possesses four broad pinkish stripes present sub-dorsally and laterally in pairs. The egg, larval and pupal stages last for 7, 26-46 and 6-11 days, respectively. The pest remains active from end of February to middle of November.

All stages of this pest are simultaneously met with front April to October showing overlapping generations. However, distinct broods are also reported in different places during June- October. Larvae hibernate from mid November to February. The fourth and fifth broods overlap during February-June. Rainfall appears to favor its multiplication. The incidence is high in heavy soils and under waterlogged or flooded conditions.

Management

- Collection of moths in light traps to minimize pest proliferation.
- Collection and destruction of egg masses and cane tops showing primary infestation.
- Conserves the parasitoid *Cotesia flavipes* and released inoculatively or augmentatively for the control of the borer.
- Release of *Trichogramma chilonis* egg parasitoids@50000/ha at 10 days interval from July is useful in reducing the infestation level.
- Avoiding indiscriminate use of insecticides and allows them to bring about natural control.

10.2. Sucking Pest

10.2.1. Sugarcane Leaf Hopper, *Pyrilla perpusilla* (O-Homoptera: F-Lophopldae)

In India, among the sucking pests of sugarcane leaf hopper is the most destructive. It is widely distributed in the oriental region. Subsequently, its epidemics have been observed in subtropical sugarcane growing tracts at intervals of 5-6 years.

Damage Symptoms

- Adults and nymphs suck cell sap from under surface of leaves and leaves turn pale yellow later shrivel up
- They also excrete honeydew on the foliage leading to sooty mould development which is reducing photosynthesis.

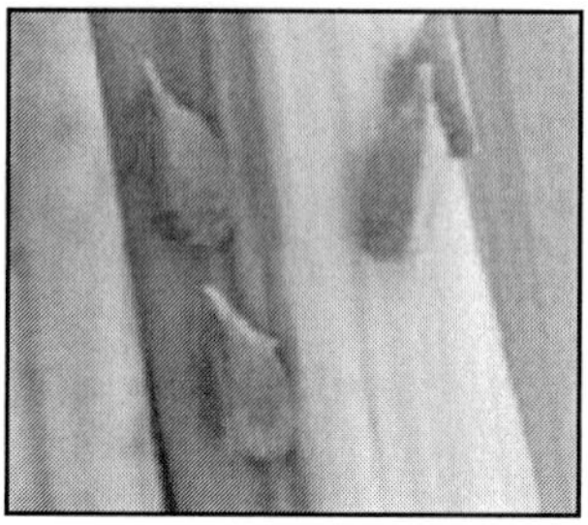

Figure 14. Adult of Pyrilla

Life Cycle

It is a straw colored insect with a tapering head. Its nymphs are tiny white hoppers with two fluffy tails. Both adults & nymphs suck the juice of cane leaves & turn them brown when present in large numbers. Infested leaves develop sooty mould, eggs are laid in clusters covered with white fluff, on the under surface of the leaves till about June-July & inside leaf sheaths later in the season. The active period of their development is from March to Early November.

Management

- Clean cultivation.

- Proper disposal of trash after harvesting to be utilized as compost or burnt.
- Detrashing of leaves from August onwards reduces pyrilla populations.
- Removal and destruction of egg-bearing sugarcane leaves at regular intervals to minimize buildup of populations.
- Avoid Ratooning where field is heavily infested.
- Collection & Destruction of eggs & insects of all stages as far as practicable.
- Conservation of the egg parasite *Tetrastichus pyrillae* and *Epiricania melanoleuca* nymph and adult parasite in the fields.
- Collect egg masses at regular interval and kept in old cloth bags and suspended at different places in the field to facilitate selective emergence of adult egg parasitoids.
- Spraying of dimethoate 30 EC @ 0.1% in detrashed crop to combat emergency build up of the pest.

10.2.2. Scale Insect, *Melanaspis glomerata*

Its attack to be very severe in parts of north India,this pest was considered to be of minor importance unitl 1951, when it was again reported in a serious form from Coimbatore. In subsequent years, it has been observed from Bihar Tamil Nadu, Gujarat, Maharashtra, Uttar Pradesh, Madhya Pradesh, Andhra Pradesh, Delhi, Karnataka and Harayana. The pest is also invading new cane areas of the country, where its occurrence was not known in earlier years. Currently, scale insect epidemics are of common occurrence in several states of the country, posing serious threat to sugarcane production.

Nature of symptoms and extent of damage

- They feed on parenchyma tissue of stem and prevent the accumulation of sucrose in the cane.
- If heavy populations of the pest cover the entire cane, the encrustation giving a grayish black appearance and the crop loses general vigor, shrivels up and exhibits stunted growth with reduction in internodal length.

Figure 15. Damage symptom & scale insect

Biology and life cycle

Adult female lays 270-400 elongate, narrow and ovoid eggs which are rounded at the anterior and posterior ends. They are light yellow in color and gradually turn to pinkish red as the embryonic development advances. After a very short egg period of 4-12 h, the nymphs or crawlers hatch within the body of the mother and escape through the genital aperture. Female passes through five instars with no difference between the fifth instars and the adult except for the size and outer scale. The scale encrustation is thick, round, grey to light black in color and convex with white waxy coating on the outer surface. The adult female is oval in shape and yellow in color. The male insect passes through four nymphal instars and a pupal stage before reaching the adult stage. Reproduction is generally by fertilization and occasionally by parthenogenesis. The life cycle of female varies from 39-49 days in summer and 50-57 days in winter. Male insect completes its development in a shorter period of 18-22 days. In the field, all stages of the insect are generally observed throughout the year indicating that the generations overlap. High temperature and high humidity prevailing during July-September favor population buildup. Well spread but non-beating rains coupled with high humidity from September onwards promote the growth of scale populations. Rainwater and high velocity winds disperse crawlers to new areas.

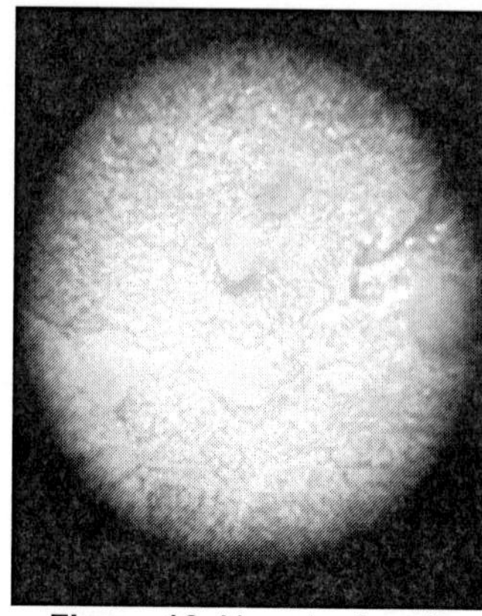

Figure 16. Nymph of scale insect

Management

- Avoidance of movement of susceptible varieties from infested areas to new areas.
- Selection of healthy planting material that should be free from insect.
- Prevention of water logging reduces humidity and helps minimize scale populations.
- Soaking setts in imidacloprid 17.8 SL @ 0.5/litre of water for 15 minutes.
- Detrashing of crop two or three times after internode formation.
- Propping of cane facilitate insecticide application in endemic areas.
- Avoiding ratoons where heively infested area for two to three years would prevent rapid multiplication and perpetuation of the pest.
- Application of dimethoate 30 EC @ 0.1% after detrashing would enhance the effectiveness of insecticide.

10.2.3. Mealy Bug, *Saccharicoccus sacchari* (O-Homeptera:F- Pseudococcidae)

The mealy bugs are ubiquitous in all the sugarcane growing tracts or the world, but seldom assume a major pest status. Because of the presence of a whitish, powdery filamentous mealy coating on the body of these insects, they are commonly referred to as 'mealy bugs'.

Damage Symptoms

- The mealy bugs derive their name from the whitish, powdery filamentous mealy coating on their body.
- They are found under sheathing leaves in colonies and suck up the juice.
- Severe infestation results in stunting of canes and yellowing of leaves.

Biology and life cycle

The pink mealy bug *Saccharicoccus sacchari* is the most commonly occurring species throughout the sugarcane growing areas of the country. The adult females are distinctly pink in color, with a round or dumb-bell shaped, clearly segmented body. The body of adult female is soft and legs are moderately long and not hairy. The black ant often attends the mealy bug. Adult females are flattened and covered by a white mealy secretion. Gregarious in habit like the pink mealy bug, they are pale yellowish or light pink in color.

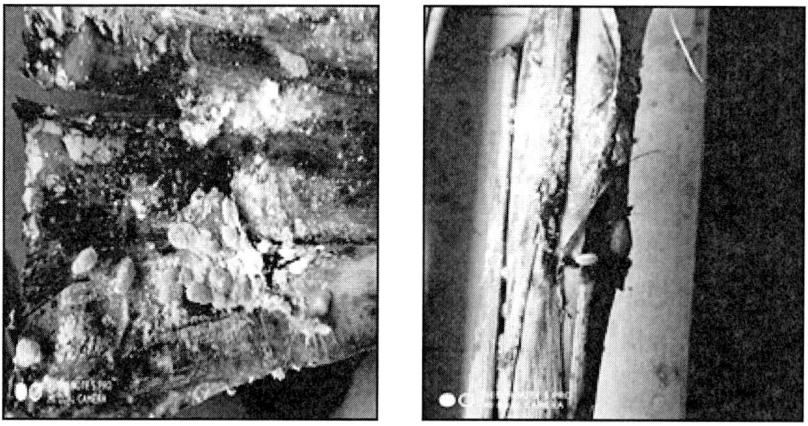

Figure 17. Damage symptoms of mealybug

The adult females are elongate, oval and small sized with a waxy white secretion covering the body. There are short waxy filaments along the margins of the body. Legs are long and slender. Drought situation is a major factor that influences mealy bug populations. They are often found in large numbers on stunted canes. Varieties with tight clasping leaf sheaths less incidence of mealy bug populations compare to lose clasping leaf sheaths.

Management

- Planting uninfested setts with leaf sheaths removed reduces perpetuation of mealyhugs to new areas as well as new plantings.
- Detrashing is essential to expose the colonies to abiotic (high temperature and sunlight) and biotic (natural enemies) stresses.
- Avoidance of overdose of nitrogen.
- Avoiding ratoons in prone areas, ratoons can act as the inoculums for the next season.
- Management of water stress and general health of the crop is important to reduce mealy bug incidence.
- In cases of severe infestation, insecticides dimethoate 30EC @ 2ml per lit should be sprayed after detrashing of cane.

10.2.4. White-Fly, *Aleurolobus barodensis* (O-Hemiptera:F-Aleyrodidae)

Two species of white-fly, viz. *Aleurolobus barodensis* Mask, and *Neomaskelia bergii Sign*. Occur as sugarcane pests. *A. Barodensis* is more common species particularly in the north Indian sugarcane belt. Its attack is generally associated with poor crops, particularly ratoons which usually remain neglected. The build-up of the pest population is rapid under water-logged conditions and in soils deficient in nitrogen. Drought also during summer and dry spells during monsoon profoundly influence whitefly buildup.

Damage Symptoms

- The nymphs of white flies suck the sap from the undersurface of the leaves.
- As a result, the leaves turn yellow and pinkish in severe cases of infestation and gradually dry up.
- Heavily infested leaves are covered by the sooty mould, (*Capnodium* sp.,) which adversely affects photosynthesis.
- Yellow streaks appear on the infested leaves and the crop acquired a yellowish-green appearance.
- Ratoon crops are more susceptible to this pest.

Figure 18. Symptom of white fly incidence

Life Cycle

Eggs are laid in rows. Nymphs who are sometimes termed puparia are in the beginning black oval specks fixed on leaves but as they grow they secrete white waxy coating over their bodies. Adults are minute insects with two pairs of wings. The egg, larval and pupal periods, besides longevity of adult flies and the total life occupies about 2-3, 5-7, 7-10, 2-3 and 15-25 days respectively. The pest passes through 9-12 generations in the course of a year.

Management

- The crop should be properly looked after and adequately manured with a nitrogenous fertilizer. In lowlying areas where water-logging is common, rationing of the crop should be avoided.

- Detreshing and destruction of leaves bearing whitefly puparia before the emergence of the adult which can be determined by the presence of "T" shaped split in the dorsum of the puparia.
- In the initial stages insecticide sprays with dimethoate 30 EC @ 0.1% can be restricted to border areas.

10.2.5. Black Bug, *Cavelerius excavates* (O-Hemiptera:F-Lygaidae)

Two species of black-bug (i) *Blissus gibbus* F. and (ii) *Macropes (cavelerius) excavates* Dist., are reported to infest sugarcane crop in north India. The latter species, M. excavates commonly known as the lygaeid bug is a serious pest of ratoons in Punjab, Haryana and Western Uttar Pradesh.

Damage Symptoms

Figure 19. Black bug

- The nymphs and adults are found in the leaf whorls and under the sheathing bases of leaves.
- Both nymphs and adults suck the sap resulting in stunting of racoon crops.
- The leaves of affected plants turn pak yellow with brown patches.
- In severe infestation,the tips and margins of the leaves dry up gradually and the entire plant may wither.
- The growth of the plant is arrested and the quality of juke is adversely affected
- Black bug incidence is heavy in ratoon fields. Where trash is wintering individuals are responsible for the outbreak of the pest during the pre-monsoon period.
- The loss due to this pest damage is severe in unirrigated fields.

LIFE CYCLE

M. excavates passes winter (October-March) as adults. They remain in the soil round the sugarcane stumps up to a depth of 5-8 cm, feeding and laying eggs. The nymphs appear in March-April; from June to the end of October the pest is present in all stages.

The adults live 3 to 4 months. Mating takes place after about a week of emergence. In summer the eggs are laid in clusters on the inner side of sheathing passes of the side-leaves and in winter singly (occasionally in pairs) close to the roots of the plants in the soil, the number of eggs laid by a female may vary from 50 to about 500.The egg-stage varied from 9-17 days during May-September. They passes 5 nyphal instars. Total duration of nymphal stage is 5 to 7 weeks in March-July and 4 to 6 weeks in July-October. The pest passes through three generations in a year.

Management

- To clean sugarcane stubble from the fields during March-April when the pest is usually found in huge numbers in fresh ratoon sprouts, proves helpful in destroying the pest in the early stages of its activity.
- The trash and the dry leaves left in the field after harvesting the crop can be utilized for this purpose.
- Application of the insecticide should be target in whorls to reach that particular portion of the plant. Application of dimethoate 30 EC @ 2 ml/litre of water directed towards the leaf.

10.2.6. Sugarcane Wooly Aphid, *Ceratovacuna lanigera* (O-Hemiptera: F- Aphididae)

Damage Symptoms

- The attack generally begins in the middle leaves and extends to upper leaves.
- Pest can identify from a distance by the presence of white patches of plants contrasting with the green canopy background.
- The wax filaments that detach from the body of thousands of aphid drop down and cover the ground with a powdery coating giving it a white appearance which easily catches the attention of one's eye upon entering the field.
- The indirect damage includes the development of sooty mould on the honeydew deposits which interferes with photosynthesis; infested cane tops are believed to affect the health of cattle when fed to them.

Life Cycle

Winged females generally initiate the attack in a new crop by producing pale greenish nymphs. With a short life span of 7-9 days, alates produce 8-43 young days giving birth to 34-62 pale yellowish nymphs whose developmental period is 6-22 days. These apteous forms multiply profusely producing large colonies of thousands of individuals on the undersurface of leaves protected from direct sunlight. While winged forms are responsible for plot-to-plot or long distance dispersal, the active first instar nymphs disperse to adjacent leaves or plants in the contiguous canopy in the morning hours. Another form of dispersal is through congregations of young nymphs on the leaf tip which are blown off by wind in a process called ballooning. The first and second instar nymphs are naked, i.e. without the white woolly filaments; dorsal side of the body of the third and fourth instar nymphs produces waxy filaments.

The last instar nymphs and wingless adult females are densely covered with thick, white cotton like secretion on the dorsum. The aphids have a pair of characteristic frontal horns and the tubular cornicles are reduced to pore like structures. Under mild tropical conditions of peninsular India where the pest occurred in epidemic form, the aphid could potentially multiply throughout the year including summer. However, June-October appeared to be the most preferred season in many studies, provided the suitable 5-6 month old crop is available in June for initiation of aphid colonies.

Management

- Complete avoidance of transport of seed material or green tops as fodder from affected areas to minimize the risk of introduction into a new area.
- Destruction of affected leaves in the early stages of infestation and priority for harvest of affected cane to prevent spread of the pest in recently invaded areas.
- Detrashing enhances accessibility for monitoring and application of insecticides.
- Avoidance of excessive and late application of nitrogen, and excess irrigation to minimize severity of attack.
- Wider spacing and paired row planting, and wrapping and propping operations render the crop canopy more amenable to monitoring and chemical control operations.
- Temporary replacement of the popular cultivars to avoid proliferation of the aphid. The popular varieties can be brought back into cultivation after the aphid populations are stabilized by the use of natural enemies.
- Intensive weekly monitoring of the field is essential due to the high multiplication rates and dispersal ability of the aphid.
- Spraying of insecticide dimethoate 30 EC @ 2 lit./ha.

10.3. Subterrenian Pest

10.3.1. Termite, *Odontotermes obesus* (O-Isoptera:F-Termitidae)

Termites are considered to be major pests in several parts of the country. The termites damage setts, shoots, canes as well as stubbles. Termite infestation occurs soon after planting and continues till harvest.

Damage Symptoms

- Termite infestation occurs soon after planting when germinating setts and young shoots are affected.

Figure 20. Damage symptom

- Newly planted setts suffer the most serious damage as termites enter through their cut ends or through the buds and feed on soft tissues replacing them with soil.

- The infested stools or shoots dry up soon after germination and these can be pulled out easily.
- Such shoots show no compensatory tillering.
- Occasionally 40-60% of the eye buds are destroyed leading to poor germination, gaps in the field and subsequent yield loss.

Life Cycle

Males and females mate during the nuptial flight, shed the wings and search for a suitable site for founding a colony. The first batch of eggs laid by the female hatches into workers which take care of expansion activities of the nest. Subsequently, soldiers are produced which protect the colony and foraging workers. Queen termite grows unusually large, up to 3-4 inches in length and about an inch in diameter, and becomes an egg-laying machine depositing thousands of eggs every day for several years. Workers forage for food and, hence, are responsible for crop damage almost round the year. Since workers cover long distances inside their protective earthen tubular walkways, colonies need to be located some distance away from the site of crop damage for mechanical destruction. Termite activity is observed in both loamy and heavy soils; areas of low rainfall or with poor irrigation potential are known to be more prone to its damage.

Management

- Location and destruction of termite colonies.
- Ploughing or disturbing high field bunds would expose and destroy colonies of subterranean termites.
- Clearing and burning of crop residues and other debries in the field.
- Frequent irrigation would help in reducing attack.
- Dipping setts for 2 min in solution of imidachoprid 17.8 SL @ 5 ml or chlorpyriphos 20 EC @ 50 ml per 10 lit of water prior to planting.
- When termite attack appears in standing crop, then chlorpyriphos or imidachloprid at the above dosages for effective control.

10.3.2. White Grub, *Holotrichia consanguinea* (O- Coleoptera: F- Scarabaeidae)

While grub termed generally used for larvae of Scarabaeid beetles. The grubs have C shaped, fleshy, white or grey coloured body. This species has widely distributed among all sugarcane belts and it is an important group of soil pests of sugarcane in India

Damage Symptoms

- Grubs feed on main roots of sugarcane and damage the underground portions of stalks.
- Due to feeding, leaves can turn yellow and such clumps gradually dry up and easily pulled out.

- Affected canes collapse, dry up and become unfit for crushing or planting.
- The damage generally occurs in patches.
- During April-May, the peak infestation is observed in August.
- The pest overwinters in adult and grub stages in earthen cells from October to February.
- Pupation occurs in March and adult emergence takes place in the April-June.

Life Cycle

Eggs are laid in the soil in sugarcane fields, fallow lands or grassy bunds along water channels. First instar grubs hatching after 10-12 days feed on roots of organic matter initially and switch to sugarcane roots. After a short one month period in each of the first two instars, third instar grubs feed voraciously on sugarcane roots for three to four months. Pupation takes place in earthen cells from September and the adults emerging in 10-14 days remain in the soil until early summer showers in the following year stimulate them to emerge out of the soil.High moisture is detrimental to the grub mid heavy rains and waterlogged conditions form; the grubs to come to soil surface. Although deforestation is responsible for its extensive spread, factors such as monocultures.

Management

- The mass emergence of beetles after the fast heavy summer shower easy for their large scale hand collection of insect.
- During the peak period of adult emergence light trap may be use for collections of adult.
- Heavily infested fields may be ploughed with tractor and the grubs, pupae and beetles hand-picked and destroyed.
- Repeated ploughing reduces the grubs' populations.
- Altering the time of planting.
- Paddling and crop rotation should be done with paddy.
- Frequent hoeing and stirring up of soil near the base of tender shoots reduce attack
- Ploughing and flooding in February when planting is taken up during the pest activity period would be useful in minimizing grub populations.
- *Beauveria brongnianii* formulated with carrier materials such as press mud, lignite or talc may be applied at 2.5 x 10'' spores / ha during June-July.
- The formulation may be mixed with suitable quantity of farm yard manure or well cured press mud and distributed in the harrows followed by irrigation.
- Apply Chlothinnidin 50% WDG @ 0.05 kg a.i./ha.

10.4. Defoliators

The defoliators are of sporadic pests in sugarcane and gene rally considered to be minor pests. This category includes army worms grass hoppers etc. They are polyphagous insects, infesting almost all crops.

10.4.1. Armyworm, (Noctuidae: Lepidoptera)

The army worm was first reported to be a pest of sugarcane in Bombay (Waste. 1951) and later in Bihar (Butani, 1955). It has assumed importance as a sugarcane pest in other states as well. Besides, it also creates havoc in paddy, wheat, maize and other monocotyledonous plants.

Damage symptoms

- The larvae chew the leaf lamina either partially or completely and leaving only the midrib.
- They take shelter beneath the leaf-sheaths or dry trash during the day and feed voraciously during nights.

Life Cycle

Eggs are laid in rows or clusters either between the overlapping leaf-sheath or on soil mostly at night. Each egg mass consists of 20-76. The egg is shiny white or light brown in colour when fresh and gradually turns pale yellow and finally black. The egg stage lasts for 5-19 days depending on weather conditions. Neonate larvae are active, dull white in colour and later turn green. The full grown larva is cylindrical in shape and four clear longitudinal stripes on either side of the mid-dorsal line of the body. The young larvae spin threads and hang on to the host plants. They feed on tender tissues of the whorl by scraping. Pupation takes place generally near water channels. In the pre-pupal stage, the larvae become yellowish brown and spin a cocoon in the soil at a depth of 0.5-5.0 cm. The pre-pupal star lasts for 1.2 days. Newly formed pupae are dirty white to straw coloured but soon change to reddish brown and finally to dark brown.

Management

- Burning of trash and cane stubble.
- Ploughing of heavily infested fields to expose the larvae and pupae for predation by birds, and predators.
- Collection and destruction of larvae and pupae.

10.4.2 : Grass Hoppers *Hieroglyphus banian* (Acrididae: Orthoptera)

Damage Symptoms

The hoppers of the first and second instar are gregarious in nature and feed for about a week on *Cynodon dactylon* in and around the fields. In the third instar stage, they migrate to sugarcane and continue to feed on leaves to the extent of 57 per cent. When the infestation is severe, defoliation may be complete, leaving only the midribs.

Figure 21. Damage symptoms of Grass hopper

Life Cycle

Eggs are laid 24 hours after mating. The time required for oviposition varies from 80450 minutes. The number of eggs per swami varies from 68.99 with an average of 81. The eggs are yellowish brown in colour. These are clustered in sub cylindrical pods with varying size. Eggs hatch in the follow ing year after the commencement of rains. Egg period ranges from 10-11 months. The eggs may exceptionally remain viable for 10 years. Six moults are noticed exclusive of the vermiform stage. The total nymph period hats for 48.81 days. Adults are green with three transverse dark lines on the pro-thorax.

Management

10.5. Other Pests

10.5.1. Mites (Acarina : Arachnida)

Four mites are common in India. They are the red mite, Oligonyehus Indian, yellow mite, *Oligonychus sacchari*, rust mite, Tarsonemor and Eriophyid mite (blister mite/ sheath mite). Mites are generally considered to be minor pests, though occasionally they explode to very high populations in large areas. The mites infesting sugarcane usually colonies on the leaf lamina except eriophyid mites which infest the leaf sheath.

Biology

Egg is tiny roundish, bead like and pale green in colour. Eggs are laid tingly in rows inside the web. One web contains 15 or more eggs. Number of eggs laid by a female varies from 21 – 48 at the rate of 2-11 eggs per day. Freshly hatched nymph is spherical. Translucent and yellowish. Subsequently, it elongates and turns green with dark specks on the dorsum. There are three pairs of legs. Soon after hatching, the nymphs are sluggish and remain within the web, but later move out and 'pin new webs and moult there in.The second miter nymph. Known as protonymph has four pain of legs. It is yellowish soon after moulting, but changes later to greenish while. In male, the abdominal tip it wedge-shaped. while in the female it is round.It is also yellowish soon after moulting, but turns to light green later on. The sexes arc distinguished by the caudal extremity. Which is narrow in males and comparatitely broad and round in females.

Management

- Removal and destruction of plants infested are recommended for suppressing its population build-up.
- Keeping the periphery of fields clear of wild host plants is also recommended. Frequent irrigation in the pre-monsoon period reduces infestation

Chapter

11

DISEASE MANAGEMENT

Disease is one of the important constraints in sugarcane cultivation. Over dozens of diseases caused by fungi, bacteria, Viruses & mycoplasma depending upon the area in which they infect the crop are responsible for this sizable decrease in terms of both total produce & productivity of sugarcane. The major diseases of sugarcane commonly occurring in the country & exerting huge losses are Red rot, smut, wilt, Grassy shoot disease mosaic; ratoon stunting, leaf spots etc.

11.1. Fungal disease

11.1.1. Red rot, *Colletotrichum falcatum*

In India this disease was first recorded by Barber in year 1901 at the Godawari Delta of Andhra Pradesh. Presently this disease is of wide occurrence in varying degree where ever the sugarcane crop is under cultivation in the country & particularly rampant in eastern U.P & northern Bihar making these places as hot spots. The disease is most devastating nature has made several promising sugarcane varieties obsolete. Diseased seed setts are the main source of survival & spread of the pathogen.

Symptoms

The initial symptoms appear as third fourth leaves from the top start drying from margin inwards & ultimately entire leaves dry and Mid-rib lesions become conspicuous during monsoon. The lesions usually start as minute red spots on the upper surface of the mid-rib & further develop forming long lesions during later stage.The canes become shriveled and lighter in weight. When the canes are split open longitudinally the pith is found reddened accompanied by white traverse patches at right angles. In advance stage of the disease the red colour may be replaced by dirty brown & white bands look hazy or unclear. These dried canes often emit sour odour & Juice does not set well on boiling.

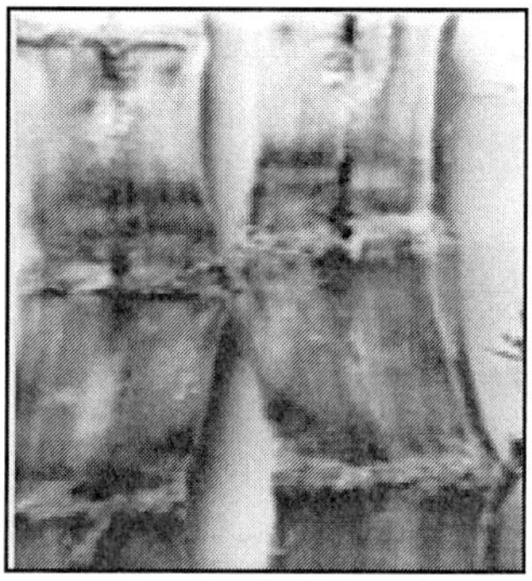

Figure 22. Damage symptom of red rot

Management

- Crop rotation should followed with rice and green manure crops.
- Select the setts should be disease free.
- Aviod ratooning when field heavily infested from diseased.
- Soak the setts in 0.1% Carbendazim solution for 15 minutes before planting.
- Grow resistant varieties.
- Setts can be treated with aerated steam at 52° C for 4 to 5 hours and by moist hot air at 54° C for 2 hours.

11.1.2. Wilt, *Cephalosporium sacchari*

Earlier wilt disease was confined to northern belt of sugarcane particularly Bihar state from where it spread to U.P., Punjab & Haryana. It is transmitted mainly through infected seed setts. The fungus can also survive in the soil. This disease commonly occurs with the infection of red rot & poor crop is more prone to wilt infection. Disease intensity largely depends upon faulty drainage & prevailing draught or in sufficient moisture conditions in the field.

Symptoms

The most striking symptoms of the disease become apparent late in the season which are yellowing accompanied by drooping of the top when the crop is ready for harvest, the growth of plants is held up & the affected canes dry rapidly. In the initial stage of the disease when the canes are split open, the tissues, particularly of the lowest internodes have a brick-red / dirty red colour in the form of conical shaped spots. This reddening may be confined to a few internodes or extended to the entire length of the cane. Such canes dry up, become hollow & there is considerable reduction in the quality of juice. Diseased canes produce characteristic foul odour.

Figure 23. Damage symptom of wilt

Management

- Select the planting material from the disease free fields.
- Avoid the practice of ratooning in diseased fields.
- Burn the trashes and stubbles in the field.
- Grow coriander or mustard as a companion crop in the early stages of crop.
- Dip the setts in Carbendazim for 15 minutes.

11.1.3. Smut, *Ustilago scitaminea*

The fungus transmitted through infected seed setts and can also survive through the spores fallen on the ground.

Symptoms

Affected plants are characterized by the production of a long, whip like structure with black dusty mass of spores at apex. During early stage this structure is straight & becomes curved, several feet in length subsequently. The whip remains covered with a bright membrane. On bursting of membrane the spores come out & get healthy plants infected by air. Such structures can be observed in flushes; first during the months of May-June and secondly during October -November. Infected plants often have slender & thin canes with reduced growth, become lighter in weight due to drying of canes & sometimes many tillers develop.

Figure 24. Symptoms of smut

Management

- Plant material should be taken from disease free area.
- Remove and destroy the smutted clump.
- Collect the whips infected part safely in a polythene bag and kill the spores to prevent further spread.
- Avoids ratooning of the diseased crops having more than 10 per cent infection.
- Crop rotation should be follow with green manure crops or dry fallowing.
- Grow red gram as a companion crop between 2 rows of sugarcane.
- Grow resistant varieties recommended for the area
- Burn the trash/ infected plant debris after crop harvest.

11.1.4. Sett Rot, *Ceratocystis paradoxa*

The fungus survives in the soil and in the infected cane. The inoculums move from field to field through wind-borne or irrigation or rain water. Inside the sett it spreads rapidly through the parenchymatous tissues and causes sett rot and also known as pineapple disease.

Symptoms

The primary symptom of disease appears on setts after 20 to 30 days of planting. The fungus is soil-borne and enters through cut ends of seed pieces and multiplies rapidly in the parenchymatous tissues. At initial stage, the affected tissues first develop a reddish colour which turns to brownish black in the later stages. The heavily infected setts show internodal cavities covered with the mycelium and abundant spores. Affected stalk when split, a characteristic pineapple smell is associated with the rotting tissues. Rotting in affected setts drastically

Figure 25. Damage symptoms of sett rot

reduce due to presence of inacculam. Poor germination due to setts may decay before the buds germinate or the shoots may die after reaching a height of about 6 to 12 cm and infected shoots are stunted.

Management

- Setts treatment with 0.05% Carbendazim and dip for 15 minutes.
- 3 to 4 buds setts may used.
- Provide drainage facilities during rainy seasons

11.1.5. Pokkah Boeng, *Fusarium moniliforme*

This fungus disease is found almost all sugarcane growing area of India and disease appears during periods of hot humid conditions. It is a wind-borne fungus and primarily transmitted through the air-currents and secondary transmission is through the infected setts, irrigation water, splashed rains and soil. The name, pokkah boeng, is derived from a Javanese word meaning a malformed or distorted top.

Symptoms

Unique symptoms of this disease are a distortion and shortening, wrinkling and twisted of the top and leaves, yellowing at the base of young leaves. In severe cases, mature stalk of cane look like a knife cut into the side and such cases yield losses can occur.

Figure 26. Symptoms of Pokkah boeng

Management

- Direct control of this disease is usually not required because the cane usually recovers with minimal yield loss.
- Some varieties are more resistant than others, but all commercial varieties have sufficient resistance to avoid significant yield loss in most situations.
- If severe condition Copper Oxycloride 50 WP may be applied @1lit.lha

11.2. Bacterial disease

11.2.1. Leaf Scald, *Xanthomonas albilaneans*

Symptoms

First white stripes appear on leaves of infected plant. These stripes are run full length of the leaf blade and these stripes are narrow and located over veins. in later stage these stripes resulting in death of the leaves lateral buds sprouted and shortening of internodes.

11.2.2. Ratoon Stunting Disease, *Leifsonia xyli*

Coryneform bacterium is mainly responsible for causing Ratoon stunting disease (RSD) in sugarcane which spreads through disease setts.

Symptoms

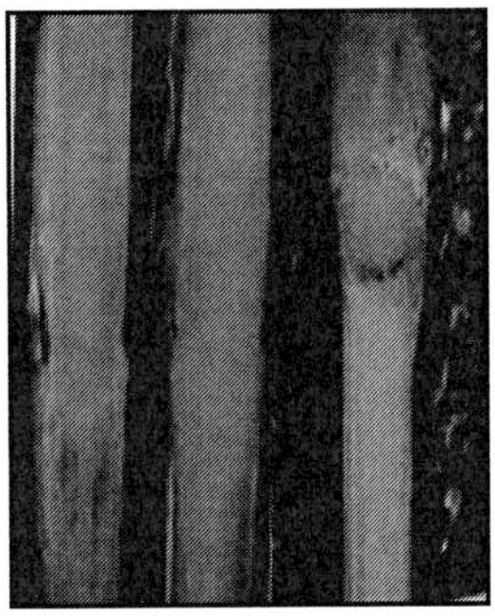
Figure 27. Damage symptoms of RSD

Yellowish leaves, reduced tillering, Stunted internodes and dwarfing of the plant are the Characteristic symptoms of RSD. The infected canes when split open longitudinally orange red vascular bundles in shades of pink, red & reddish brown or yellow orange at the nodes can be seen. Well defined symptoms appear in the crop deficit in moisture, nutrients etc.

Management

- Setts should be select from disease free fields.
- Remove disease infected clumps and burn it.
- Setts treatment should be done before planting.

11.2.3. Red Stripe , *Pseudomonas rubrilineans*

The inoculums of this pathogen remain viable in the soil and infected plant residues. The bacterium primarily spreads through infected canes. The secondary spread is mainly through rain water, irrigation water and may be through insects. Infected parenchymatous cells may collapse and normal functioning of the plant parts may fail. Continuous ratooning and prolonged rainy weather with low temperature are most favourable for this disease.

Symptoms

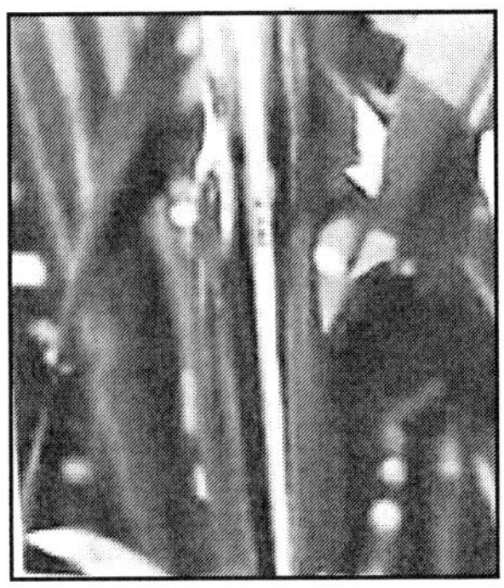
Figure 28. Damage symptoms of Red stripe

The disease first appears on the basal part of the young leaves. The stripes appear as water soaked, long, narrow chlorotic streaks on leaves and turned reddish brown in few days. These stripes are 0.5 to 1 mm in width and 5-100 mm in length and run parallel to the midrib. The stripes mostly confined to lower half of the leaf lamina. The whitish flakes spreads to growing points of the shoot and develop yellowish stripes that later turn reddish brown. The rotting may started from the tip of the shoot and spreads downwards. The core is discoloured to reddish brown and shriveled and form cavity in the centre. In badly affected fields, a foul smell appears.

Management

- When the disease is noticed in the field, the affected plants should be removed and burnt.
- Growing resistant varieties as per recommendation of that areas
- Disease free setts should be selected.
- Avoid to growing alternate hosts near the sugarcane fields.

11.3. Viral disease

11.3.1. Sugarcane Mosaic Disease

In India the disease was first observed by Barber in 1927. Now its appearance has been reported from different sugarcane growing areas in the country. Infection of this disease occurs through infected seed setts. Aphid, Rhopalosiphum maidis is the main vector for secondary infection.

Symptoms

The disease symptoms characteristically appear on basal portion of foliage than on the older leaves prominently in the form of yellowish or chlorotic stripes alternate with green space of the leaf a mosaic pattern. Considerable increase in chlorotic area over the normal green & appearance of symptoms on the leaf sheath become common features during advance & severe infections. Yellow stripes also appear on the rind of the internodes & stalks finally dry up forming 'sunken' areas called as canker stage of mosaic. The disease affects both quantum of sugar & Jaggery as well as their quality.

Management

- Roguing of infected plants and use healthy planting material.
- In early crop stage sprays insecticides Imidacloprid 17.8 SC@ 500ml/ha to manage the insect vector population.
- G row resistant cultivars.
- Rogue out the diseased clumps periodically.
- Select setts from disease free fields.
- Aerated Steam Therapy at 56° C for 3 hrs of setts before planting.

11.3.2. Sugarcane yellow leaf disease

It is viral disease and transmitted by aphids, *Melanaphis sacchari* and *Rhopalosiphum maidis*, this virus is also spread by infected planting material. Dry weather conditions during October to until the end of harvest is favorable condition of this disease.

Symptoms

Yellowing of the leaf midrib on the underside of the leaf is a characteristic symptom of this disease. Initial symptoms of yellow leaf, with a yellowing of the lower surface of the leaf midrib of leaves 3 to 6 counting from the top expanding spindle leaf. Yellowing is most prevalent and noticeable in mature cane from October until the end of harvest. The yellowing expands out from the leaf midrib into the leaf blade as the season progresses until a general yellowing of the leaves can be observed from a distance.

Figure 29. Symptom of yellow leaf disease.

Management

- Use disease free planting materials
- Yellow sticky trap use for control of this vector
- Apply Imidacloprid 17.8SL @ 500ml/ha twice at 15 days of interval.

11.4. Phytoplasma

Grassy Shoot Disease

The GSD was first observed by Barber (1919) & reported by vasudeva (1955) from Belapur, Maharashtra. The disease has been recorded in most of the sugarcane growing belt in India since then but the incidence being particularly high in Maharashtra.

Symptoms

The disease is characterized by proliferation of auxiliary buds from base of the cane giving rise to profuse/ crowded bunch of tillers bearing narrow leaves which exhibit varying degrees of loss of chlorophyll, ranging from total green to white (albinism). Cane formation rarely takes place in affected clumps & if formed they are thin with short internodes giving plant a bushy appearance. The primary transmission of disease is through infected cane setts while the secondary transmission is through insect vectors i.e. aphids *Rhopalosiphum maidis, Melanaphis sacchari & M indosacchari.*

Figure 30. Symptoms of GSD

Management

- Eradicate diseased infected plant as symptoms are appearing.
- Selection of setts should be avoided from diseased area.
- Pre-treating the healthy setts with hot water at 52° C for 1 hour before planting
- Treating them with hot air at 54° C for 8 hours.
- Spraying the crop twice a month with insecticides

11.5 Integrated Disease Management

11.5.1 Soil management

- Inoculums of diseases viable in soil and plant debris so that all cop residues should be burnt to reduce the inoculums of these diseases.
- During summer season deep tillage should be done to expose the inoculums to direct sunlight and due to high temperature inoculums will die.

11.5.2 Seed management

- Replacement of susceptible varieties to resistance varieties.
- Use seed material from a crop free from diseases.
- Both cut ends of setts should examine properly if end of setts hollow and red area should be rejected.

11.5.3 Seed treatment

- Chemical treatment is used for protection of cut ends to avoid entry of soil borne pathogens like Sett rot, red rot, smut and wilt through cut ends.
- Indofill M-45 @ 0.25% or Bavistin at 0.1% as sett dip for 10 minutes is recommended.
- This will not only protect the cut ends from soil borne pathogens but also enhance sett germinability by 10-15%.
- Setts are also treated with heat to inactivate or eliminate the pathogens.

  Hot-water 50 °C -2 hours

  Moist-hot-air 54 °C- 4 hour

11.5.4 Crop management

- Season and date of planting have a significant influence on disease development.
- Autumn planted cane crop showed resistance against diseases because its susceptible phase passes at time (March-June) when climatic conditions are unfavorable for pathogen. Spring planted crop is vulnerable to attack as the crop passes through the susceptible phase in (July-October).
- Monoculture of sugarcane over large areas provides congenial conditions for the rapid buildup of inoculums.
- As companion crop Garlic, Onion, Coriander, Linseed, and Mustard help to reduce inoculums loads of soils.
- Spraying of crop All though, several fungicides have been found effective against sett rot, top rot, root rot, leaf spots
- Rotation with rice may be helpful to reduce wilt inoculums.
- Flooding and ill drained conditions help in the spread of inoculums of red rot, smut, wilt and sett rot diseases. To avoid this, upland should be selected as far as possible; failing which good drainage should be ensured for mitigating the diseases.
- Irrigation water should not flow from diseased filed to disease free field.
- Frequent diseased clumps should be removed from field to prevent further spred of disease.

11.5.5 Regulatory Measures

- In India no regulatory measures are imposed during interstate or zonal movement of seed materials so that quarantine rules are strictly followed to check movement of pathogen one place to other.

11.5.6 Biological Control

- Seed, soil treatment and foliar spraying with *Trichoderma viride* and *Trichoderma harzianum* helps to reduce disease incidence

❖❖❖

Chapter

12

DIFFERENT PRACTICES ADOPTED IN SUGARCANE

12. Different practices adopted in sugarcane

12.1. Earthing up and intercultural operations

(a) Earthing up

Earthing up should be done 2-3times during crop period in sugarcane. The first earthing-up is known as "partial earthing-up and to cover the fertilizers and provide anchorage to root system. The second operation is "full earthing-up". Full earthing-up is done after final manuring (90-120 days coinciding with peak tillering). The soil from the ridge is thrown on both sides towards cane rows and the furrows will become as ridges and ridges as furrows. The furrows so formed are used for irrigation.

(b) Wet earthing up

- Done around 6 months age of the crop
- The furrows are irrigated and the wet soil from furrows is taken and plaster the ridges
- It checks late tillering and watery shoots
- Heavy earthing up is useful during floods
- When the flood water recedes, the excess water from earthed-up soil drains out quickly thus providing aeration

12.2. Hoeing

This operation is done where soil crust formation is very common. In sub-tropics hoeing is done after germination is over using a bullock drawn or a tractor-drawn harrows. While carrying out this operation some of the germinated setts may be uprooted and they are pressed down manually.

12.3. Removal of water shoots

Late-formed tillers or side shoots that grow robustly and rapidly are known as water shoots. They originate mostly from excess water supply, heavy and late manuring, insufficient earthing up. These shoots contain plenty of water, low sucrose and more sugar reduction. Water shoots affect the growth of stalks adjacent to them. They harbour insect pests and are low in recovery when milled sugar recoveries are low because of reduced juice quality. Therefore, removal of water shoots whenever they appear is necessary water shoots can be used as cattle feed.

12.4. Control of flowering

Flowering is not desirable in commercial sugarcane cultivation. The cane growth stops once the plant has flowered and starts to mature. If not harvested immediately reversion of sugar, increase in fibre, pith formation, cane breakage, etc. may occur. The deterioration is much faster if it is summer. Growers can be used non-flowering or shy flowering varieties where flowering is a serious problem, with controlled irrigation and use of growth regulating substances. Spraying of ethrel @ 500ppm, twice or 1000ppm once at floral initiation or before.

12.5. Propping

Tying the canes by using the lower bottom leaves to check lodging of cane. Propping can be either done for each row or two rows can be brought together and tied. Main advantages of propping are to prevent the crop from lodging. It is extensively followed in coastal belt where cyclone effect is very severe. Lodging is very common in tall varieties with less fibre content.

12.6. Wrapping and Tying

By wrapping the distribution of CO_2 become easy and proper to all the plants throughout the field. The dried leaves are removed from the plants and the green leaves on plants are wrapped together by taking all the canes of one bundle. After wrapping the clumps in adjacent rows are tied together (crosswise). When wrapping and tying is over in entire field tied clumps are given further support of bamboo poles from outside (on two opposite sides) of the field.

12.7. Detrashing

Detrashing can be done by hand operated or mechanically operated sugarcane detrasher. Detrasher removes the leaves. It also removes the sprouted buds. This is a simple hand operated device that helps farmers to detrash the sugarcane leaves. To detrash the leaves of sugarcane is a big task.

Figure 31: Different practices in sugarcane like tying, wrapping, staking

Chapter

13

SUGARCANE HARVESTING

13.1. Maturity and harvesting

Ripening of sugarcane refers to rapid synthesis and storage of sucrose in the stalk. Accumulation of sugar in the stalk starts soon after completion of elongation phase. Glucose produced during photosynthesis is not utilized for conversion but stored as sucrose. When the concentration exceeds 16% in the juice and 85% purity the cane is said to be matured.

13.1.1. Criteria for harvesting of Sugarcane

There are some criteria of maturity of sugarcane for harvesting are as follws:

1. **Age basis:** Harvesting is done based on maturity (age) group.
2. **Sign and symptoms:** Yellowing and drying of leaves, metallic sound of mature canes when tapped appearance of sugar crystal glistening when a mature cane is cut in a slanting way and held against the sun are some of the visual indices of assessing maturity of cane.
3. **Quality parameters:** Cane maturity can be assessing on the basis of quality parameters of sugarcane which are as followed:
    (a) **Juice Brix:** Juice Brix refers to the total solids content present in the juice expressed in percentage. Brix includes sugars as well as non-sugars. Brix can be measured in the field itself in the standing cane crop using a Hand Refractometer or Hand Refractometer Brix or HR Brix. Brix percentage 16-18 % in sugarcane juice confirms that it is ready to harvest.

    (b) **Juice Sucrose or Pol Per Cent:** The juice sucrose per cent is the actual cane sugar present in the juice. It is determined by using a polarimeter, hence sucrose per cent is also referred to as pol per cent. For all practical purposes pol % and sucrose % are synonyms now a day's an instrument called sucrolyser is also available for determining sucrose % in juice.

(c) **Purity Coefficient:** It refers to the percentage of sucrose present in the total solids content in the juice. A higher purity indicates the presence of higher sucrose content out of the total solids present in juice. The purity percentage along with sucrose percent aids in determining maturity time.

Purity Percentage = (Sucrose %/HR Brix) x100

A cane crop is considered fit for harvesting if it has attained a minimum of 16% sucrose and 85% purity.

4. **Reducing Sugars:** The reducing sugars refer to the percentage of other sugars (fructose and glucose) in the juice. A lower reducing sugars value indicates that much of the sugars have been converted into sucrose.

5. **Commercial Cane Sugar:** The commercial cane sugar (CCS) refers to the total recoverable sugar percent in the cane. This could be calculated by the following formula:

   CCS (tons/ha) = [Yield (tons/ha) x Sugar Recovery (%)] /100

   Sugar Recovery (%) = [S - 0.4 (B - S)] x 0.73

   Where, S= Sucrose % in juice and B= Corrected Brix (%)

6. Fibre % in cane weight of bagasse produced from the weighed cane crushed. It can be determined by using following formula.

   Weight of Bagasse = (weight of cane + weight of added water) – weight of the mixed juice obtained.

7. **Top : Bottom Ratio (TBR) :** It refers to the ascending or descending order of sugar (sucrose) accumulation in cane. A cane with0.95 to 1.0 TBR is ready for harwest & cane with 1.1 to 1.2 is over matured. Higher the TBR values, there is higher degree of deterioration.

8. **Maturity factor (MF) :**

   MF = {(ERS;, Bottom) – ERS, Top) : ERS, Bottom} × 100}

   Where ERS is, Estimated Recoverable sugars.

9. **Glucose ratio :** Fraction of reducing sugars % cane over Pol % cane. The GR ratio is smaller, during ripening.

13.2. Yield

Yield of sugarcane varies from region to region. In subtropical region sugarcane crop yield varies from 600-1000 quintal per hectare as compared to tropical region is about 800-1500 quintal per hectare depending upon the agro-climatic conditions, soil conditions and other management practices of the region.

Chapter

14

RATOON MANAGEMENT PRACTICES IN SUGARCANE

14.1. Ratoon

The word ratoon seems to have originated either from the Latin words retonsus - cut down or mown; or retono - to thunder back, resound; the Spanish - retono fresh shoot or sprout or even the French rejeton - sucker or shoot, scion, descendant, offspring or sprout. The first harvest of a crop is usually called the "plant crop" and each succeeding harvest is designated "first ratoon," "second ratoon," and so on.

14.1.1. Trash burning

Trash burning is essential only under the following conditions: 1. Plant crop infected by pests and diseases like scales, mealy bugs, etc. 2. Areas prone to heavy termite attack 3. Areas prone to rodent attack 4. Excess moisture affecting sprouting 5. Soils where sub soil drainage is poor, and 6. Areas where fire hazard exists.

14.1.2. Stubble shaving

After harvest of the standing crop and cleaning the field, an indispensable operation that has to be carried out is "stubble shaving". The stubbles protruding out of the field are cut below ground level using sharp spade. This is done to facilitate healthy underground buds to sprout and establish a deeper root system in the ratoon crop. Generally, buds above ground get damaged during harvesting and subsequent cleaning of the field. Besides, buds would be dry and would have been infected with disease causing organisms. Thus, if the above ground buds are allowed to germinate, sprouting would be inadequate to establish a good crop stand. Besides, the fresh shoot-roots also would not be able to enter the hard soil thus affecting absorption of nutrients and water. This is why many a times drying up of shoots occur inspite of having fairly adequate sprouting. Successive ratoons originate from a higher point. Thus, the root system in the successive ratoons is progressively raised which is undesirable. By deep stubble shaving, the lower buds are forced to germinate and thus have root system at lower level. Deep root system thus obtained would help utilise nutrients and moisture from the lower soil layers. Deep root system is also necessary to give good anchorage to the ratoon crop. For stubble shaving, a sharp implement, preferably a spade should be used so that stubbles are cut with minimum damage to the buds and stubbles are not uprooted. If the soil has become hard and dry and it is difficult to carry out these operations, the field should be irrigated lightly ad stubble shaving may be done when field conditions are ideal for working with implements. In some varieties, as in the case of CoC 671, the root system is shallow and there is problem of uprooting of stubbles. In such cases instead of a spade, it is better to use a sharp knife to cut the stubbles to the ground level. Tractor drawn stubble shaver at the Indian Institute of Sugarcane Research, Lucknow, has been developed and satisfactorily tested for shaving the stumps close to the ground level. Drive for the cutter unit is taken from the P.T.O. of the tractor and is transmitted through universal joint and bevel gear arrangements. The implement gives an out-turn of 2 ha per day and brings down the cost of stubble shaving by 50 per cent.

14.1.3. Off-barring

It is an operation wherein the ridges are broken or cut on either side. This operation is also called as "shoulder breaking". To loosen the soil, the inter spaces between the rows are dug. Shoulder breaking can also be carried out by bullock drawn implements like wooden plough, or small ridger like implement. Sub soilers may also be used for breaking the compacted soil. But for passing a sub-soiler, a heavy duty tractor is required. These operations reduce soil compaction and facilitate quicker development of fresh root system and help in vigorous growth of the young crops.

14.1.4. Gap filling

One serious problem in ratoons is the occurrence of gaps, which when exceeding 20 per cent cause considerable yield losses. Gaps occur because of poor sprouting owing to several reasons: cold or hot weather conditions, poor plant crops, attack of fungal diseases, insect pests etc. A spot in a row can be considered as a gap if for a distance of about 60 cm there is no cane clump. Otherwise, there is no need for gap filling. For gap filling, it is better to use pre-germinated single bud setts. For this purpose, a month prior to harvest of the plant crop, nursery may be planted with single bud setts and seedlings if required for gap filling can also be obtained from spots where excess sprouting is seen. Clumps can be operated and cut into quarters and planted in the gaps. Gap filled seedlings or quartered clumps require more attention. Pot watering may have to be ensured for better establishment. In delta region, quartered clump transplanting in the gaps has been found quite effective in the wet land soils. Polybag seedlings for gap filling another recent technique is to raise polybag seedlings and use them for gap filling. For this purpose, single bud setts are planted in polythene bags (measuring 12 x 8 cm having a few punch holes for aeration) filled with soil, sand and FYM in the ratio 1:1:1. One month old seedlings can be used for gap filling. While gap filling, the leaves of the seedlings should be clipped off. Then after removing the polythene bag, the seedlings along with the soil mass should be placed in a small pit made in the gap. A small quantity of super phosphate placed in the pit helps better growth of the seedling. In this method, 100% of the gap filled seedlings will survive.

14.1.5. Other agronomic management

Agronomic practices like fertilizer and manure, water management, weed management and harvesting should be practiced according to the main sugarcane crop.

14.2. Advantages of ratoon cropping

Reduced cost of production through saving in land preparation and care of the plant during early growth of the clone; such early growth is eliminated because of the presence of a readily formed root system. Ratoons are economical by about 25-30 per cent in the operational cost because of saving in the cost of setts and initial preparatory cultivation. Better utilization of growing season, especially in monsoon climates. Higher yield per unit area in a given period of time. Ratoon crop can be used in breeding to maintain the same plant of a clone through several seasons. Often uses less irrigation water and fertilizer than main crop because of the shorter growth period. Ratoons save time as they establish early and in general mature early. Therefore they can be harvested early. Farmers will get time to grow other crops or have enterprises in the farm. Ratoons stabilize the cane area of a factory. Each year, the factory will have assured cane area to the extent of ratooning, which is around 40-50 per cent in most areas and therefore every year the effort for planting is limited to the fifty per cent of the total cane area required.

Chapter

15

MECHANIZATION IN SUGARCANE

15. Implements and machinery used for sugarcane cultivation

15.1. Sugarcane planter

Cane planting includes Planting of two rows of sugar cane. It opens the furrows in 75 cm spacing and cuts the cane into 37 cm long setts. Then placed the setts in the furrows along with dropping of fertilizer below the setts and dispensing the insecticide solution over the setts. After planting it covers the setts with soil and providing light compaction to minimize sett and soil moisture loss. Advantages of this implement is to save Irrigation water, Seed and Fertilizer up to 20- 30%, 40% and 40% respectively.

15.2. Tractor operated pit digger for sugarcane planting

It dugs two pits of 90 cm diameter simultaneously at 1.5 m interval to a depth of 30 cm suitable for planting sugarcane setts. Planting of cane in 1.5 x 1.5 m spacing with pit method favors higher cane yield. Recommended as technology package under drip fertigation system for the cane. Its function is to dug pits.

15.3. Sugarcane sett cutter

Sugarcane sett cutter cuts the sugarcane setts with a single bud.

15.4. Sugarcane detrasher

It removes the sprouted buds and collects the de-trashed leaves of sugarcane. This detrasher is requiring less number of labour.

15.5. Sugarcane Crusher

Extract the of juice of sugarcane

15.6. Sugarcane Inter-cultivator

This newly developed inter-cultivator has two sets of tynes with three tynes in each set. This can be hitched to the heel-adjusted tractor. It can be run on the field 40 days after planting to remove the weeds in between the crop in two rows. The escaped weeds and other large dicot weeds within the rows can be removed by engaging minimum labor.

15.7. Earthing up ridger

Its main function is to make ridges and earthing up the sugarcane field. Earthing up ridger is an implement, which can be run in between the crop rows to cover the nutrients applied as a first top dressing on the 40th day of planting. When the crop is nearing 80 to 90 days of age, a second top dressing is done, and the same implementation can be used again for full earthing up of the plants. The furrows are uniformly deep, and it facilitates the uniform flow of irrigation water in the field.

15.8. Sugarcane stubble shaver

It is a single row stubble shaving with 0.26 ha of field capacity. It helps in better ratoon emergence. It is a sugarcane stubble shaver driven by a tractor. The innovation can cut a single row in a pass whereas the swinging blade type shavers can cut two rows in a pass but it will give the superficial cut.

15.9. Ratoon manager

This machine is designed for stubble shaving, off-baring and harrowing. Fertilizer & pesticide applicator may also be available as an option, if needed. Ratoon manager will be the best if the field preparation (Ridges and Furrow) is done with a tractor and suitable for different row spacing. It uniformly shaves the stubbles and helps to cut the roots by the off-baring. Harrowing facilitates better aeration and water penetration in order to rejuvenate and induce the growth of the sugarcane.

15.10. Ridger cum weeder

The ridger-weeder is peculiar because it also does weeding as well as earthing-up in a single sugarcane crop activity. Cost and time savings is more than 60 per cent. Therefore the implementation can be used very effectively in crop rows of 1.5 m standard spacing without disturbing the crop. The soil is cut to a depth of 22.5 cm and is raised on either side of the wingboard. The raised soil is thrown on both sides. n the field, the tractor is driven in 'alternating field machine pattern' by moving in one direction from one end of the field between the first and second line. Depending upon the type of soil the machine will cover about 0.8 ha per hour.

15.11. Rotavator

Unlike conventional implements, a seedbed with one or two passes can be formed by Rotavator. The combination tractor / rotavator may begin cultivation much earlier than usually drawn implements after rain. It is most necessary for the reduction of Sugarcane stubble and the eradication of rhizomes.

15.12. Sugarcane bud chipper

It Remove the buds from a cane. It is floor-mounted and designed to surgically cut out the buds with a knife with a semicircular edge. The machine helps the user to sit comfortably on the ground and feed the cane continuously with the left hand, while swinging the right arm in a smooth arc using the ergonomic spring-loaded handle to cut the sugarcane buds.

15.13. Foot pump sprayer

The foot sprayer is one of the perfect and reliable sprayers that are used to spray multipurpose work. The operating concept is identical to the rocker sprayer. The sprayer consists of a foot lever pump, a strainer suction hose, a delivery pipe, a spray lance mounted with a shut-off gun valve, a gooseneck bend and customizable nozzles. The pump barrel is mounted on steel frames which provide it stability when placed on the ground. It has a provision of two strong springs which retract the foot lever to its original position after each pumping stroke.

15.14. Solar powered manual sprayer

In order to spray herbicide / fungicide on various crops, a solar powered manual sprayer has been developed. The height of the spray can be changed from 300 mm to 1000 mm. Preliminary testing of this sprayer for pre-emergence herbicide spraying in sugarcane has been conducted at the IISR, Lucknow Institute Farm.

Figure 32. Picture of IISR developed solar powered manual sprayer

15.15. Sugarcane combine harvester:

It is acceptable for the fields of Plain, Mountain, Hill, Dry regions. It has the purpose of loading sugar cane, wood and other agricultural uses. This combines harvester cuts the sugar cane into pieces, but it is very expensive. The combined harvester of sugarcane will harvest @ 1.2 ha / day, that is, 20 tonnes / hr.

Figure 33 : Different implements used in sugarcane cultivation

Chapter

16

NUMERICAL PROBLEM RELATED TO SUGARCANE

16.1. Calculation of seed rate in sugarcane

Example: Calculate the seed rate of sugarcane for 1 acre land

Average No. of nodes on the cane = 35

Setts to be used for planting = 3 budded setts

Average no. of damaged buds on each cane = 5

Average weight of cane = 980

Average length of each 3 budded setts = 23 cm

Spacing: 120 cm end to end, 90 cm bud to bud

Solution: Total number of setts in end to end system of planting

$$= \frac{10000 \times 1}{1.2 \times 0.23}$$

$$= 36231.8$$

Total number of canes needed for one hectare

$$= \frac{\text{No. of nodes/sett} \times \text{Total No. of setts/ha}}{\text{tota no. of nodes per cane} - \text{damaged nodes/cane}}$$

$$= \frac{3 \times 36231.8}{35\text{-}5} = \mathbf{3623.18}$$

Weight of cane needed for planting one hectare area

= Total number of canes needed x Average wt. of cane

$$= \frac{3623.18 \times 980}{1000 \times 100} \text{ quintals/hectare}$$

$$= 35.50 \text{ quintals/hectare}$$

In case of bud to bud system of planting usually each sett is overlapped to the extent of 2/3rd of its length. Therefore, the total number of setts needed would be = 3/2 × setts needed /ha for end to end system of planting.

Thus, the total numbers of setts would be

$$= \frac{3 \times 36231.8}{2}$$

$$= 54346.5/1000 = 54.34 \text{ quintals/hectare}$$

Example: A farmer is having land of about 0.4 ha want to grow sugarcane in the field through end-end fashion. Calculate the number of setts required when the length of internode of the 3 budded setts is 25 cm and row to row spacing is 90 cm.

Solution: Length of internode of sett = 25 cm = 0.25 m

Row to row spacing is 90 cm = 0.9 cm

Total area required for1 sett = 0.25 × 0.9 sq.m = 0.225

So, 0.225 m^2 area is for one sett

10000 m^2 area required for $\frac{1}{0.225} \times 10000$

= 44,444 setts

16.2. Fertilizer calculation

Example: Let the recommended dose of fertilizer dose for sugarcane crop be 150, 80, 50 kg Nitrogen, Phosphorus and Potassium per hectare respectively. The amount of fertilizer required in the form of urea, Single superphosphate (SSP) and muriate of potash (MOP) is calculated as shown:

Urea containing about 46% Nitrogen, Superphosphate about 16 % Phosphorus and 60 % Potassium.

Formula: The amount of fertilizer to be applied

$$= \frac{100}{\text{nutrient percentage in fertilizer}} \times \text{recommended dose}$$

46 kg N supplied from 100 kg urea

150 kg N will be supplied from $\frac{100}{46} \times 150 = 326$ kg urea is required per hectare

Likewise

16 kg P_2O_5 supplied from 100 kg SSP

80 kg P_2O_5 will be supplied from $\frac{100}{16}$ x 80 = 500 kg SSP is required per hectare

Similarly

60 kg K_2O supplied from 100 kg MOP

60 kg K_2O will be supplied from $\frac{100}{60}$ x 60 =100 kg MOP is required per hectare to supply 60 kg K_2O

Example: Assume that N, P_2O_5, K_2O are to be applied for sugarcane at the rate of 150:60:40 kg/ha respectively. Nitrogen is to be applied in 5 splits, full phosphorus and potash at the time of planting. Find out the quantities of urea, SSP and MOP needed for basal and top dressing of fertilizers.

N application rate is 150 kg/ha, it should be applied in 5 equal splits, it means we have to apply 150/5 = 30 kg N at each splits at different growth interval.

46 kg Nitrogen obtained from 100 kg urea

30 kg Nitrogen will be obtained from = (100/46) x 30

= 65.22 kg urea

Amount of SSP to supply 60 kg P_2O_5

16 kg P_2O_5 per hectare obtained from 100 kg SSP

60 kg P_2O_5 obtained from (100/16) x 60 = 375 kg SSP

Calculation of MOP for supply of 40 kg K_2O

60 kg K_2O per hectare obtained from 100 kg MOP

40 kg K_2O obtained from (100/60) x 40 = 66.66 kg MOP

16.3. Calculation on water requirement

Example : 100 ha crop area has to be irrigated by a pump working for 20 ha day. Crops are to be irrigated at 50% DASM. Water holding capacity of soil is 20 cm m-1 depth. Mean depth of the root zone is 75 cm. conveyance and application efficiencies are 75 and 80% respectively. Average consumptive use rate of crops is 4 mm day-1. Calculate the following:

- Net irrigation requirement?
- Gross irrigation requirement?
- Required capacity of the irrigation system.

Solution : WHC of the soil $= \frac{20 \times 75}{100} = 15$ cm

Net irrigation requirement (crops are irrigated at 50% DASM)

$$= 15 \times \frac{50}{100} = 7.5 \text{ cm}$$

Gross irrigation requirement

$$= \frac{\text{Net depth of irrigation (cm)}}{\text{Conveyance efficiency (\%)} \times \text{Application efficiency (\%)}}$$

$$= \frac{7.5}{0.75 \times 0.80} = 12.5 \text{ cm}$$

$$\text{Irrigation period} = \frac{\text{Net depth of irrigation (cm)}}{\text{Average daily consumptive use}}$$

$$= \frac{75}{4} = 23.5 \text{ Days}$$

Required Capacity of irrigation system

$$= \frac{\text{Depth of irrigation (cm)}}{\text{Irrigation period}}$$

$$= \frac{12.5 \times 50\text{ha} - \text{cm}}{23.5 \times 20 \times 60 \times 60\text{s}}$$

$$= \frac{12.5 \times 50 \times 100000}{16{,}92{,}000 \text{ s}} = 369.38\text{lps}$$

Example : Calculate cumulative evaporation required for scheduling irrigation at 0.75 and 0.8 with 6 cm of irrigation water?

Solution: Cumulative pan evaporation at IW/CPE ratio of 0.75 = IW/CPE = 0.75

$$\frac{6}{\text{CPE}} = 0.75, \text{CPE} \times 0.75 = 6$$

$$\text{CPE} = \frac{6}{0.75} = 8 \text{ cm}$$

Irrigation of 6 cm is given when CPE is 8 cm

CPE at 0.8 ratio = 6/0.8 = 7.5 cm

In IW/CPE ratio approach, irrigation can also be scheduled at fixed level of CPE by varying amount of irrigation water.

Example : Calculate the amount of water for each irrigation for scheduling irrigation at 0.6 and 0.9 IW/CPE with 12 cm of CPE?

Solution : Amount of water to be given at IW/CPE ratio of 0.6 = IW/12 = 0.6

IW = 0.6 12 = 7.2 cm

Amount of water to be given at IW/CPE ratio of 0.9 = IW/12 =0.9,

IW = 12 0.9 = 10.8 cm

Example : Soil is sampled 3 days after irrigation at 20 cm depth and again when the crop depleted the soil moisture to PWP (the crop died) with the following results. the plants. The furrows are uniformly deep, and it facilitates the uniform flow of irrigation water in the field.

Values	Three days after irrigation (FC)	Soil moisture at PWP
Volume (cm_3)	121	144
Wet weight (g)	197	215
Dry weight (g)	158	193

Determine the FC and PWP as per cent volumes, ASM and the bulk density?

Solution : Soil moisture content at FC = 197 – 158 = 39 g or 39 cm^3

$$\text{FC on volume basis} = \frac{39}{121} = 32.23\%$$

Soil moisture content at PWP = 215 – 193 = 22 g or 20 cm^3

$$\text{PWP on volume basis} = \frac{22}{144} = 15.28\%$$

Available soil moisture (ASM) = 32.23 – 15.28 = 16.95% = 17 cm m^{-1}

$$\text{Bulk density at FC} = \frac{158}{121} = 1.31 \text{ g cm}^{-3}$$

$$\text{Bulk density at PWP} = \frac{193}{144} = 1.34 \text{ g cm}^{-3}$$

Example: Calculate the water holding capacity of 60 cm root zone soil of sugarcane field from the following data

Soil depth (cm)	Filed capacity (%)	PWP (%)	Bulk density (g cm^{-3})
0-15	22.7	10.5	1.47
15-30	22.6	10.4	1.51
30-45	22.8	10.9	1.53
45-60	22.4	10.7	1.49

Solution : M_{si} - M_{hi}

$$\text{Water holding capacity of soil} = \sum_{i=1}^{n} \times \frac{M_{si} - M_{hi}}{100} \times A_{si} \times Di$$

- For 0-15 cm depth $= \dfrac{22.5 - 10.5}{100} \times 1.47 \times 15 = 2.69$ cm

- For 15-30 cm depth $= \dfrac{22.6 - 10.4}{100} \times 1.51 \times 15 = 2.76$ cm

- For 30-45 cm depth $= \frac{22.8 - 10.9}{100} \times 1.53 \times 15 = 2.73$ cm
- For 45-60 cm depth $= \frac{22.4 - 10.7}{100} \times 1.49 \times 15 = 2.62$ cm

Water holding capacity of 60 cm root zone

$$= 2.69 + 2.76 + 2.73 + 2.62 = 10.8 \text{ cm}$$

16.4. Calculation of herbicides

Formula: Amount of herbicide (kg or litre)

$$= \frac{\text{Rate desired (kg a.i / ha)} \times \text{Area (ha)} \times 100}{\text{Herbicide concentration (\%)}}$$

Example: Calculate the amount of Atrazine 80 % required covering an area of 5 ha where the rate of application is 2.0 kg a.i /ha. Also calculate the amount of water to be added in herbicides when the rate of water required 500 litres per hectare.

Solution : $Q = \frac{10 \times R \times A}{P}$

Where, Q = Quantity of herbicide in grams or millilitres

R = Rate of application in kg a.i/ha

A = Area in m^2

P = Percentage of active ingradient (a.i) in the chemical

$$Q = \frac{10 \times 2 \times 5 \times 10000}{80}$$

= 12500 g or ml = 12.5 litre for 5 hectare area

Therefore, the amount of water required

$$= \frac{500 \times 5 \times 10000}{10000}$$

= 2500 litre water for spraying 5 hectare area

16.5 Calculation of pesticides

1. **If pesticides recommended for farmer as kg a.i. /ha:**

Formula : Rate of pesticides is given mainly in terms of a.i. /ha

$$\text{Quantity of material required per hectare} = \frac{\text{Rate of application}}{\text{Active ingredient in \%}} \times 100$$

Example. Find out the quantity of Carbofuran 3G to be applied in one hectare area if rate of application is 1 kg a.i. /ha

Quantity of Carbofuran /ha = $\frac{1}{3} \times 100 = 33.3$ kg/ha

- For the calculation of this type of numerical a.i. mention in the commercial product.

2. **If pesticides recommended for farmer as % concentration:**

Formula :

$$\text{Amount of pesticide} = \frac{\text{Volume of spray solution (liter)} \times \text{Percent strength of pesticide solution to be sprayed}}{\text{Percent strength of pesticide given (a.i./1 or kg)}}$$

Example. Amount of chlorpyriphos 20 EC when sprayed as 0.1 per cent solution

$$= \frac{500 \times 0.1}{20} = 2.5 \text{ liter or } 2500 \text{ ml/ ha}$$

16.6. Estimation of yield of sugarcane

Example: Estimate the yield of sugarcane, juice, sugar and Jaggery from the following data

1. Average number of millible canes/ clump = 18
2. Average weight of cane = 700 g
3. Spacing = 120 cm × 45 cm
4. Extractable juice in sugarcane = 75 percent
5. Sugar percentage in juice = 18
6. Percentage jaggery in juice = 25

Solution: Yield of sugarcane in tonnes per hectare

$$= \frac{\text{Average wt. of canes in g} \times \text{No. of milliable canes / clump}}{1000 \times 100 \times 10 \times \text{spacing of clump in sq.m}}$$

$$= \frac{700 \times 10000 \times 18}{1000 \times 100 \times 10 \times 1.2 \times 0.45} = \mathbf{233.3\ t/ha}$$

The total amount of expected juice in canes

$$= \frac{233.3 \times 75}{100} = 175 \text{ t/ha}$$

The total amount of expected cane sugar

$$= 175 \times \frac{18}{100} = 31.5 \text{ t/ha}$$

The total amount of expected jaggery

$$= 175 \times \frac{25}{100} = 35 \text{ t/ha}$$

References:

2017. 3rd Advance estimates for sugar season: 49 (7)

Ambika, S., Somasundaram G. and Sundaralingam, K. 2015. Importance of Green and Green Leaf Manures in Sustainable Agriculture. Popular Kheti: 3(1): 62-65

CACP. 2013. Price Policy for Kharif Crops 2013-14. Commission on Agricultural Costs and Prices. Department of Agriculture & Cooperation, Government of India. pp 112.

CWC. 2014. Guidelines for Improving Water Use Efficiency in Irrigation, Domestic and Industrial Sectors. Ministry of Water Resources, Central Water Commission. Government of India.

CWC. 2015. Water and Related Statistics, Central Water Commission. Water Resources Information System Directorate, Water Planning and Project Wing. Government of India.

Das, T. K. 2009. Weed Science- Principles and Application. Jain Publishers.

Gupta, O. P. 2007. Weed Managemen-Principles and Practices. Agrobios.

https://sugarcane.dac.gov.in/pdf/StatisticsAPY.pdf

https://www.kisansuvidha.com/sugarcane/?v=ad4f1670f142

IWM in sugarcane, Agropedia

Rana, S. S. and Rana, M. C. 2016. Principles and Practices of Weed Management. Department of Agronomy, College of Agriculture, CSK Himachal Pradesh Krishi Vishvavidyalaya, Palampur, 138 pages.

Rao, V. S. 2000. Principles of Weed Science. Oxford & IBH.

TERMINOLOGY

1. **Bagasse:** It is fibrous residue left over after sugarcane is crushed. It is known as "Mill Wet Bagasse because of high moisture content (50 %). It contains nearly 46-52 % fibre, 2.5-30. % Sugar and 0.55 % mineral constitutes. It is also used as dry bagasse for fuel purpose.
2. **Molasses:** Heavy, viscous liquid separated from the final low grade massecuite from which no further sugar can be crystallized out by the usual method. It contains about 30 to 35 % sucrose, 12-16 % Ash, 10 to 25 % reducing sugars.
3. **Pressmud:** Solid residue of sugarcane generated after filtration of juice. During purification process, juice rises to the top and mud collects at the bottom in the sugar production. Pressmud is used for making compost as it contains high amount of nutrients which is beneficial for growth and development of plants/ crops. Approximately, 10-15 million tonnes of pressmud (filter cake) is produced as as waste material from sugar mills.
4. **Herbicide:** It is derived from two words. Herbi means "Plants" and cide means "to kill". Literally, Herbicide is known as weed killers which are used to kill the unwanted plants which are out of place.
5. **Bio-herbicide:** These are usually native pathogens cultured artificially and sprayed just like post-emergence (PoE) herbicides to kill the target weeds. Examples of Bio-herbicide are Devine, Collego, CASST, etc.
6. **Acid equivalent (a.e.):** The theoretical yield of parent acid from herbicide formulation is called as acid equivalent.
7. **Active ingradient (a.i.):** It is expressed as percentage by weight or by volume. Chemical present in commercial product which is responsible for the herbicidal activity is termed as active ingradient.

$$\text{Quantity of commercial product to be required} = \frac{\text{Recommended dose (kg or g) 1 ha}}{\text{active ingradient (a.i.)}} \times 100$$

8. **C_4 plant:** In C_4 plant, 4 carbon compounds are produced. Examples of C4 plants are Sugarcane, Maize, Sorghum. These plants avoid photorespiration by using enzyme called Phosphoenolpyruvate carboxylase (PEP).
9. **Auricle:** It is a small ear-like projection from the base of a leaf or petal.
10. **Relative Humidity (R.H.):** It is the ratio between the amount of water vapour present in the air and amount of water vapour required for saturation at a particular temperature and pressure. It is expressed as percentage.

11. **Monosaccharide:** It is the basic units of carbohydrates and known as simple sugars. The general formula is Cn H2nOn .Monosaccharide is usually soluble in water, colourless and crystalline solids. It is sugar that is not decomposable into simple sugars.
12. **Bulk density (B.D.):** It is defined as mass (weight) of a unit volume of dry soil. The bulk density of soil ranges from 1.0-1.5 g per cubic centimeter. Bulk density of coarse textured soil is generally higher than the fine textured soils.
13. **Field capacity (FC):** Water content in soil after removal of gravitational water from macropores (filled with air) and water is rest only in micropores after 2-3 days of irrigation or rainfall. It is the uppermost limit of water availability of plants.

    FC > ME in sandy soil

    FC < ME in clay soil

    FC= ME in loam soil

    Where, FC is Field capacity

    ME is moisture equivalent
14. **Emulsifiable concentrate (EC):** The active ingradient is dissolved insolvent and an emulsifier is added to it. Since, the emulsifier helps in uniform distribution of chemical in water, no stirring is necessary while spraying.
15. **Wettable powders (WP):** When herbicide is of low solubility in the water, they can be grounded into fine powder for suspension in water are termed as Wettable powders (WP). They require continuous stirring to prevent their settling and to give uniform concentration of the herbicide in the entire spray fluid.
16. **Suspension liquid (SL):** When active ingradient is solubilized in organic solvens but not soluble in water .when the product of a.i. and solvent is added to the water for spraying it forms suspension liquid. These are comparatively cheaper than emulsifiable concentrates.
17. **Granules (G):** Granules or small pellets are made with inert clays. The solution of the toxicant is sprayed on these granules in desired quantity and dried.

Index